The Ultimate Dental School Application Guide

UniAdmissions

ISBN 978-1-912557-40-0

Published by *RAR Medical Services Limited*
www.uniadmissions.co.uk
info@uniadmissions.co.uk
Tel: 0208 068 0438

The Ultimate Dental School Application Guide

Dr. Jason Briggs

Dr. Jessica Nazareth

Dr. Rohan Agarwal

UniAdmissions

About the Authors

Dr. Jason Briggs BSc (Hons) BDS

Jason studied Chemistry at the University of Liverpool where he qualified with honours in 2008 and then began his career working at one of the largest biotech companies in the world. He stayed in this industry for 3 years before following his true passion, a career in dentistry. He is a fully qualified dentist having recently qualified from the University of Sheffield. His interests lie in restorative dentistry and full mouth rehabilitation.

Jason enjoys rock climbing, bouldering and swimming in his spare time.

Dr. Jessica Nazareth BDS

Jessica is a fully qualified dentist who qualified from the University of Sheffield in 2016. She now works as an associate dentist in multiple practices in the North West of England providing both NHS and Private care. She has actively participated in campaigns for Oral Cancer awareness and National Smile Month, which educates the public on prevention and the importance of good oral health. She has a keen interest in preventative techniques and aesthetics.

Away from dentistry she enjoys baking, badminton, arts and crafts

Dr. Rohan Agarwal MB Chir MA (Hons) Cantab

Rohan is the **Director of Operations** at *UniAdmissions* and is responsible for its technical and commercial arms. He graduated from Gonville and Caius College, Cambridge and is a fully qualified doctor. Over the last five years, he has tutored hundreds of successful Oxbridge and Medical applicants. He has also authored fifty books on admissions tests and interviews.

Rohan has taught physiology to undergraduates and interviewed medical school applicants for Cambridge. He has published research on bone physiology and writes education articles for the Independent and Huffington Post. In his spare time, Rohan enjoys playing the piano and table tennis.

INTRODUCTION

Dental school is, and always has been, extremely competitive. It consistently attracts the top students from every school. Therefore dental schools have very difficult decisions to make. They have to decide who are the very best amongst a sea of excellent applicants. And their duty goes beyond simply allocating the space to the highest achievers, those who may deserve it most. Dental schools are selecting the dentists of the future – those who will look after the oral health of many generations across the course of their career.

This decision, therefore, has vast and wide-ranging consequences. Dental schools have both a moral and indeed a legal duty to properly assess applicants and choose the ones who will make the best dentists. But how do they decide, and how can you convince them that you are the one who deserves that valuable place?

The application process has multiple steps, which help dental schools gather information about applicants to help their decision making. The main components are your personal statement, admissions test and interview. Whilst this process may seem daunting, once you learn about each step and break each down into manageable chunks, you will find it much more straightforward and probably enjoyable!

This book covers each of the steps required in making a successful application, guiding you through the application process. In reading the book, you can benefit from the experience of the many successful applicants and specialist tutors who contributed to the resources. You can read the book as a whole in order to gain a perspective on the entire application process, or focus on individual sections to build specific skills.

This book contains elements from three titles in the *UniAdmissions Ultimate Guide* book series. It is designed to be an introduction to the dental school application process rather than a comprehensive guide for each component.

You are highly advised to look at each individual book for more advice, strategies and questions when it comes to actually sitting the exam. You can get free copies of all of the books below– see the back for more details.

➢ The Ultimate BMAT Guide
➢ The Ultimate UKCAT Guide
➢ BMAT Past Paper Worked Solutions

PERSONAL STATEMENTS

Your <u>Personal</u> Statement

Applying to university is both an exciting and confusing time. You will make a decision that will decide the next 5 years and potentially your entire life. Your personal statement is your chance to show the universities you apply to who you really are. The rest of the application is faceless statistics – the personal statement gives the admissions tutor the opportunity to look beyond those statistics at you as an individual: the person they may spend five years training and who may eventually become a dentist.

Some people may tell you there is a right and wrong way to write a personal statement, but this is a myth. Enclosed in this book you will find a number of different personal statement styles. Yes, there are rules of thumb that can help you along the way, but never lose sight of the fact that this is the best opportunity to tell *your* story.

How does the Process Work?

University applications are made through the online UCAS system. You can apply to four different universities (and also a fifth non-dental option if you choose). After receiving the outcomes of all applications, if you have offers you make a confirmed (i.e. first) and reserve choice. Your place at university is only guaranteed however once you have achieved the conditions set out in the conditional offer.

It's important to remember that the **same application will be sent to all of your choices**. There is therefore, little point in applying to completely unrelated subjects e.g. English for your fifth choice. Many applicants choose not to apply to a fifth university. Of the ones that do, most opt for a related scientific course such as biomedical science, human physiology or human anatomy.

The Timescale

For dentistry, be aware of an earlier deadline, **15 October** – if you don't get your application in on time it won't be considered. Remember that **schools often have an earlier internal deadline** so they can ensure punctuality and sort out their references in time. Different schools have different procedures, so it is very important that you find out what the timescale is at your school before the end of your AS year. Internal deadlines for dentistry can be as early as the beginning of September, which is only a couple of weeks after the summer break. Submitting earlier can be good- as it frees up your time to concentrate on admission test preparation, interview preparation and your A2 studies.

What are the Requirements?

➢ Maximum 4000 characters
➢ Maximum 47 lines
➢ Submitted by the early deadline – 15 October

What do Admissions Tutors look for?

1. Academic ability

This is the most obvious. Every university will have different entrance requirements for the same course titles, so make sure that you are aware of these. Some universities may have extra requirements; it is your responsibility to ensure that you meet the entry criteria for the course that you're applying to. Remember though not to waste time describing the A-levels you are taking – this is in another part of the form and not unique to you. Focus on describing your passions and how you have developed your knowledge in these areas.

2. Extra-curricular activities

Unlike in the US, the main factor in the UK for deciding between candidates for university places is their academic suitability for the course to which they have applied, and little else. Whilst extra-curricular activities can be a positive thing, it is a common mistake for students to dedicate too much of their Personal Statement to these. There is however an important place for subject-related extra-curricular activities especially those which demonstrate your manual dexterity. Therefore when you include extra-curricular activities such as music or art, it is often a good idea to focus on a specific transferrable skill you have gained from each – something like communication, teamwork or time management – that supports your application.

3. Manual dexterity

Dentistry involves working in a confined spaced with high precision and skill. Therefore you must be able to demonstrate superior fine motor skills and hand eye coordination. It is important to have a hobby, which involves working with your hands to help refine your manual dexterity and aid your application. Some dental schools ask for examples to be brought to the interviews and have even been known to ask for live drawing demonstrations, so do not lie about your skill set!

4. Passion for your subject

This is the most important part of a personal statement. This is what makes your statement personal to you, and is where you can truly be yourself, so do not hold back! Concentrate on the ways you first became interested in dentistry, how you discovered more about the profession and any particular areas of interest.

5. I have the grades, will I be accepted?

In short: not necessarily. Achieving the entrance grades required is considered to be the basic requirement for all successful applicants, and will certainly be the case for all applicants who gain interviews. If an applicant's personal statement isn't suitable, even if they meet the minimum grade criteria, they may still be rejected. Think of the grades as a baseline to ensure you have the academic ability, and all the other factors as ways of assessing who is the most suited for the course.

Application Timeline

Component	Deadline	Component	Deadline
Research Courses	June + July	Expert Checks	Mid-September
Start Brainstorming	Start of August	Submit to School	Late September
Complete 1st Draft	Mid-August	Submit to UCAS	Before 15th October
Complete Final Draft	End of August		

1) Researching Courses

This includes both online research and attending university open days. Whilst some of you reading this guide will already know that you want to apply for dentistry, some may not have decided. Course research is still very important even if you're certain you want to study dentistry. This is because the 'same' courses can **vary significantly** between universities. Some courses have a very separated pre-clinical science and clinical transition, whereas others are fully integrated where clinical experience starts from year one. Some are taught traditionally with lectures and practicals, whereas others have a much more self-directed problem-based learning (PBL) style. Do your course research and decide which style of course is best for you. As only one personal statement is sent to all universities that you apply to, it is important that you write in a way that addresses the different needs of each university you apply to.

If you cannot make it to university open days (e.g. if you are an international applicant), you can usually email a department and request a tour. If you allow plenty of time for this, quite often universities are happy to do this. Be proactive – do not sit around and expect universities to come to you and ask for your application! The worst possible thing you can do is appear to be applying to a course which you don't understand or haven't have researched.

Make sure you research the course content of courses that you interested in. Every university will produce a prospectus, which is available printed and online. This will help not only to choose the 4 universities that you should apply for, but also be aware of exactly what it is that you are applying for. Therefore you can address your personal statement to the particular style of the courses applied to, plus be ready at interview to explain your fit for each particular dental school.

2) Start Brainstorming

At this stage, you will have narrowed down your subject interests and should be certain that you're applying for dentistry (if you're not then check out our "*Ultimate Personal Statement Guide*" for other non-dental subjects).

A good way to start a thought process which will eventually lead to a personal statement is by simply listing all of your ideas, why you are interested in your course and the pros and cons between different universities. If there are particular modules which capture your interest that are common across several of your university choices, do not be afraid to include this in your personal statement. This will show not only that you have a real interest in your chosen subject, but also that you have taken the time to do your research.

3) Complete First Draft

This will not be the final personal statement that you submit. In all likelihood, your personal statement will go through multiple revisions and re-drafts before it is ready for submission. In most cases, the final statement is greatly different from the first draft.

The purpose of completing a rough draft early is so that you can spot major errors early. It is easy to go off on a tangent when writing a personal statement, with such things not being made obvious until somebody else reads it. The first draft will show the applicant which areas need more attention, what is missing and what needs to be removed altogether.

4) Re-Draft

This will probably be the first time at which you receive any real feedback on your Personal Statement. Obvious errors will be spotted, and any outrageous claims that sound good in your head, but are unclear or dubious will be obvious to the reader at this stage.

It is important to take advice from family and friends, however with a pinch of salt. Remember that the admissions tutor will be a stranger and not familiar with the applicant's personality.

5) Expert Check

This should be completed by the time you return for your final year at school/college. Once the final year has started, it is wise to get as many experts (teachers and external tutors) to read through the draft personal statement as possible

Again, you should take all advice with a pinch of salt. At the end of the day, this is your UCAS application and although your teachers' opinions are valuable, they are not the same as that of the admissions tutors. In schools that see many Medical and Dental applications, many teachers believe there is a correct 'format' to personal statements, and may look at your statement like a 'number' in the sea of applications that are processed by the school. There is no 'format' to successful personal statements, as each statement should be **personal** to you.

At schools that do not see many Dental applications, the opposite may be true. Many applicants are coerced into applying to universities and for courses, which their teachers judge them likely to be accepted for. It is your responsibility to ensure that the decisions you make are your own, and you have the conviction to follow through with your decisions.

6) Submit to School

Ideally, you will have some time off before submitting your statement for the internal UCAS deadline. This is important because it'll allow you to look at your final personal statement with a fresh perspective before submitting it. You'll also be able to spot any errors that you initially missed. You should submit your personal statement and UCAS application to your school on time for the internal deadline. This ensures that your school has enough time to complete your references.

7) Submit to UCAS

That's it! Take some time off from university applications for a few days, have some rest and remember that you still have A levels/IB exams to get through (and potentially admissions tests and interviews).

Getting Started

The personal statement is an amalgamation of all your hard work throughout both secondary school and your other extracurricular activities. It is right to be apprehensive about starting your application and so here are a few tips to get you started...

General Rules

If you meet the minimum academic requirements then it is with the personal statement that your application to university will be made or broken. With many applicants applying with identical GCSE and A-level results (if you're a gap year student) the personal statement is your chance to really stand out and let your personality shine through. As such there is no concrete formula to follow when writing the personal statement and indeed every statement is different in its own right. Therefore throughout this chapter you will find many principles for you to adopt and interpret as you see fit whilst considering a few of these introductory general rules.

Firstly: **space is extremely limited**; as previously mentioned a maximum of 4000 characters in 47 lines. Before even beginning the personal statement utilise all available space on the UCAS form. For example do not waste characters listing exam results when they can be entered in the corresponding fields in the qualifications section of the UCAS form.

Secondly: always remember **it easier to reduce the word count** than increase it with meaningful content when editing. Be aware that is not practical to perfect your personal statement in just one sitting. Instead write multiple drafts starting with one substantially exceeding the word limit but containing all your ideas. As such starting early is key to avoid later time pressure as you approach the deadline. Remember this is your opportunity to put onto paper what makes you the best and a cut above the rest – you should enjoy writing the personal statement!

Lastly and most importantly: **your statement is just one of hundreds that a tutor will read**. Tutors are only human after all and their interpretation of your personal statement can be influenced by many things. So get on their good side and always be sympathetic to the reader, make things plain and easy to read, avoid contentious subjects and never target your personal statement at one particular university (unless you're only applying there!).

When Should I Start?

Although it might sound like a cliché, the earlier you start writing, the easier you make the process. Starting early helps you in four key ways:

1) The most important reason to start early is that it is the **best way to analyse your application**. Many students start writing their personal statement then realise, for example, that they haven't done enough work experience, or that their extra reading isn't focused enough. *By starting early, you give yourself the chance to change this.* Over the summer, catch up on your weak areas to give yourself plenty to say in the final version.

2) **You give yourself more time for revisions.** You can improve your personal statement by showing it to as many people as possible to get their feedback. With an earlier start, you have more time to modify, thus improving the final result.

3) **Steadier pace.** Starting early gives you the flexibility of working at a steadier pace – perhaps just an hour or so per week. If you start later, you will have to spend much longer on it – probably some full days – reducing the time you have for the rest of your work and importantly for unwinding, too.

4) **You can finish it earlier.** If you start early, you can finish early too. This gives you time to change focus and start preparing for the UKCAT, BMAT (if needed), and for your interviews, which can sometimes start by mid-November.

What people think is best:

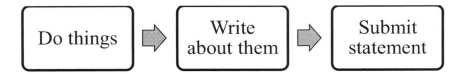

What is actually best:

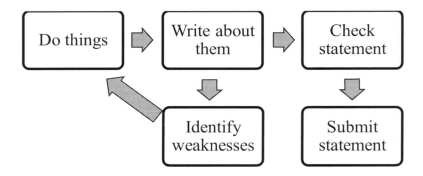

Taking your First Steps

A journey of a thousand miles starts with a single step...

As you may have already experienced, the hardest step of a big project is the first step. It's easy to *plan* to start something, but when it *actually* comes to writing the first words, what do you do? As you stare at the 47-line blank page in front of you, how can you fill it? You wonder if you've even done that many things in life. You think of something, but realise it probably isn't good enough, delete it and start over again. Sound familiar?

There is another way. The reason it is hard is because you judge your thoughts against the imagined finished product. So don't begin by writing full, perfectly polished sentences. Don't be a perfectionist. Begin with lists, spider diagrams, ideas, rambling. Just put some ideas onto paper and **write as much as possible** – it's easy to trim down afterwards if it's too long, and generally doing it this way gives the best content. Aim to improve gradually from start to finish in little steps each time.

Doing your Research

The two most important things you need to establish are: *What course?* and *What University?*

If you're unsure of where to begin, success with the personal statement begins with preparation and research.

Your choice of university is entirely personal and similar to your course choice; it needs to be somewhere that you are going to enjoy studying. Remember that where you end up will form a substantial part of your life. This could mean going to a university with a rich, active nightlife or one with strict academic prowess or perhaps one that dominates in the sporting world. In reality each university offers its own unique experience and hence the best approach is to attend as many open days as feasibly possible. At which you will have the opportunity to meet some of your potential tutors, talk to current students (who offer the most honest information) and of course tour the facilities.

The best way to prevent future stress is to start researching courses and universities early i.e. 12 months before you apply through UCAS. There is a plethora of information that is freely available online, and if you want something physical to read, you can request free prospectuses from most UK universities. It is important to remember that until you actually submit your UCAS application, **you** are in control. Universities are actively competing against one another for **your** application! When initially browsing, a good place to start is by simply listing courses and universities which interest you, and 2 pros and cons for each. You can then use this to shortlist to a handful of universities that you should then attend open days for.

There are no right choices when it comes to university choices, however there are plenty of wrong choices. You must make sure that the reasons behind your eventual choice are the right ones, and that you do not act on impulse. Whilst your personal statement should not be directed at any particular one of your universities, it should certainly be tailored to the course you are applying for.

With a course in mind and universities short listed your preparation can begin in earnest. Start by ordering **university prospectuses** or logging onto the university's subject specific websites. You should be trying to find the application requirements. Once located there will be a range of information from academic demands including work experience to personal attributes. Firstly at this point **be realistic with the GCSE results you have already achieved and your predicted A-level grades**. Also note that some universities will require a minimum number of hours of work experience – this should have been conducted through the summer after GCSE examinations and into your AS year. Work experience is not something to lie about as the university will certainly seek references to confirm your attendance. If these do not meet the minimum academic requirements a tutor will most likely not even bother reading your personal statement so don't waste a choice.

If you meet all the minimum academic requirements then focus on the other extracurricular aspects. Many prospectuses contain descriptions of ideal candidates with lists of desired personal attributes. Make a list of these for all the universities you are considering applying to. Compile a further list of your own personal attributes along with evidence that supports this claim. Then proceed to pair the points on your personal list with the corresponding requirements from your potential universities. It is important to consider extracurricular requirements from all your potential universities in the interest of forming a **rounded personal statement applicable to all institutions**.

This is a useful technique because one university may not require the same personal attributes as another. Therefore by discussing these attributes in your statement, you can demonstrate a level of ingenuity and personal reflection on the requirements of the course beyond what is listed in the prospectus.

Always remember that the role of the personal statement is to **show that you meet course requirements by using your own personal experiences as evidence**.

Brainstorming

If writing prose is too daunting, start by using our brainstorm template. Write down just three bullet-points for each of the 12 questions below and in only twenty minutes you'll be well on your way!

Why Dentistry?

What areas of dentistry interest you the most?

What are your 3 main hobbies and what skills have they developed?

What have you chosen to read outside the A-level syllabus?

Do you have any long-term career ideas/aspirations?

What did you learn from your work experience?

Have you won any prizes or awards?

What is your favourite A-level subject and why?

What are your personal strengths?

Have you attended any courses?

Have you ever held a position of responsibility?

Have you been a part of any projects?

The Writing Process

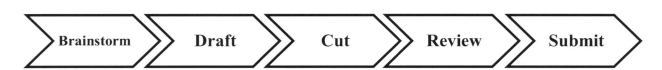

Brainstorm > Draft > Cut > Review > Submit

What is the Purpose of your Statement?

An important question to ask yourself before you begin drafting your personal statement is: how will the universities I have applied to use my personal statement? This can dramatically change how you write your personal statement. For the majority of courses that don't interview, the personal statement is directly bidding for a place on the course.

However, given that you're applying for dentistry, you will almost certainly have to go through an interview process. Therefore, your personal statement will require substantially more thought and tactics. The first thing to establish is the role of the personal statement in the context of the interview. At this point it is well worth going through the application procedures in prospectuses and on university websites.

The first option is that the personal statement is solely used for interview selection and discounted thereafter. In this case the interviewer is going to want to discuss material that isn't including in your personal statement. As such, make sure you leave yourself room to talk more at interview. Make sure you have extra material to expand on the key points in your personal statement to avoid being left in a difficult position.

Alternatively the personal statement can represent a central component of the interview. Many universities adopt an interview protocol whereby the interviewers run through the personal statement from start to finish questioning the candidate on specific points. This technique has many benefits for the interviewer as it allows them to assess the presence of any fraudulent claims (it is very hard to lie to a tutor face to face when they starting asking for specifics), it gives the interview clear structure but also allows the interviewer to bring pre planned questions on specific personal statement points.

However from the candidate's point of view this can lead to an oppressive, accusative and intense interview. There are techniques to take control back into your own hands like, for example, "planting" questions within your personal statement. This can be achieved in many ways. The phrase "for example" can be helpful here. If you describe an experience and give some examples, it gives the impression of an incomplete list, allowing you to provide extra detail at interview.

Finding the Right Balance

The balance of a personal statement can have a significant effect on the overall message it delivers. Whilst there are no strict rules, there are a few rules of thumb that can help you strike the right balance between all the important sections.

The **most important point** when applying for dentistry is dedicating enough space to talking about your work experience. This is absolutely essential – a good discussion of your work experience answers *almost all* the questions the admissions tutor will want to find out about you. **Discussing work experience can show why you've chosen dentistry, your motivation, knowledge and professional awareness.** It can also be used to highlight any specific areas of interest, hints at future career plans and to guide questions at interview. In short, it is one of the best ways to make you stand out.

Extra-curricular activities are a great way of supporting your skills. As a dental applicant you must be able to demonstrate you have the required fine motor skills to perform intricate tasks. Discussing your active participation in art or competence in playing a musical instrument can help build an admission tutors confidence in your manual dexterity. However it is generally recommended to spend no more than a quarter of the personal statement discussing extra-curricular activities, leaving the other three-quarters for discussing your motivation for dentistry, reading and work experience.

The following template gives a suggestion how to balance the different sections:

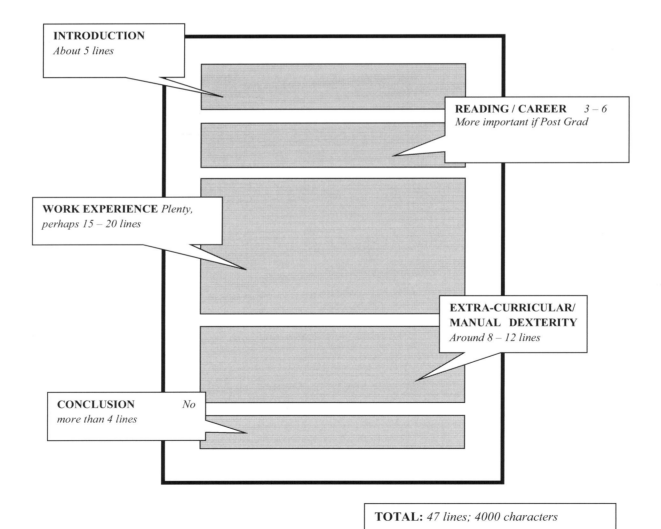

INTRODUCTION
About 5 lines

READING / CAREER   *3 – 6*
More important if Post Grad

WORK EXPERIENCE *Plenty, perhaps 15 – 20 lines*

EXTRA-CURRICULAR/ MANUAL DEXTERITY
Around 8 – 12 lines

CONCLUSION   *No more than 4 lines*

TOTAL: *47 lines; 4000 characters*

Structuring Your Statement

This may sound obvious, but many personal statements do not have a good structure. Personal statements are formal pieces of prose written with a clear purpose. Choosing a good structure helps all the important information to shine through clearly and be easily understood by the reader.

The Introduction

The Opening Sentence

Rightly or wrongly, it is likely that your personal statement will be remembered by its opening sentence. It must be something short, sharp, insightful and catch the reader's attention. Remember that admissions tutors will read several hundred personal statements and often their first impression is made by your opening sentence which is why it needs to be eye catching enough to make the tutor sit and pay particular attention to what you have written. It does indeed set the standard for the rest of the personal statement.

If this seems a daunting prospect (as it should!) then here are a few pointers to get you started:

➢ Avoid using overused words like "passionate", "deeply fascinating" and "devotion".
➢ Avoid using clichéd quotes
➢ If you are going to use a quote then put some effort into researching one that is obscure yet particular and powerful – don't forget to include a reference.
➢ Draw on your own personal experiences to produce something both original and eye-catching.

In many ways it is best that you save writing your opening statement till last, that way you can assess the tone of the rest of your work but also write something that will not be repeated elsewhere.

If you are really stuck with where to begin try writing down a memory and then explain how it has affected your relationship with your subject.

Whilst the opening statement is important, it is not something to stress about. Although a strong opening statement can make the personal statement, a bad one rarely breaks it.

Why Dentistry?

The introduction should answer the most important question of all – **why dentistry?**

*Why do **YOU** want to study dentistry?*

It is essential to show your genuine reasons and motivation. The first thing to consider is whether you genuinely want to be a dentist. You need to be certain that your motivation comes from yourself and not from external sources such as teachers or family.

Once you are certain dentistry is the right choice for you, there are a few key styles you can use. By reading the example personal statements, you will see different ways to use these building blocks to customise your own feelings towards a dental career.

➢ **Personal experience of dental treatment:** If you have had a personal experience of significant dental care and this has motivated you to study dentistry.
➢ **Family dental care:** If the dental treatment of a friend or family member has ignited your interest.
➢ **Making a difference:** If you have seen people facing difficulties or suffering from poor oral health or trauma and you would like to help in the capacity of a dentist.
➢ **Academic interest plus experience**: If your interest has spiralled out of a love of the biological sciences – but must be supported by practical experience, as work as a dentist is very different from academic science.
➢ **Suggestion plus experience:** If someone has suggested the idea of dentistry to you, made you think and investigate, and then you realised it would be a great choice for you.

Whichever reason style you choose to go with, or if you have a completely different reason altogether, a good answer always has a few key features. A good answer to the "why dentistry" question will always tell a story with three key points:

➢ When you first thought about dentistry as a career
➢ How you went about learning what the job is actually like
➢ Why you have decided it is the right choice for you

Try to avoid clichés when describing your route into dentistry. Some people will say they've wanted to be a dentist ever since they were born – but of course, this simply isn't true and therefore it isn't helpful. The admissions tutor wants to see a simple and honest story about your journey, helping them assess how carefully you have considered your choice and how suitable a choice it is.

The *exact* phrase: *"from a young age I have always been interested in"* was used more than 300 times in personal statements in 2013 (data published by UCAS), and substituting "young" for "early" gave an additional 292 statements – these phrases can quickly become boring for admissions tutors to read!

There are certain things that raise red flags, phrases that will count against you if you write them. These include: saying that dentistry is a respected career, saying that dentists are well paid, saying that you want to be a dentist because of other people telling you to, saying you want to be a dentist because other professions are worse in some way and making direct comparisons to law or engineering.

These phrases are all bad because they don't show your interest in the actual work that dentists do. Dentists may be well respected, but this fact alone won't motivate you to succeed at dental school and in a dental career – you need to be interested in what dentists *do*. The same applies to choosing dentistry because it's a stable professional occupation. Likewise, if you're choosing dentistry because someone else told you it would be a good choice (or suggested medicine or law but you don't like law or the idea of medicine), you may lack this personal motivation that dental schools know is essential for success.

The Main Body

In the rest of your text your aim should be to demonstrate your suitability for the course by exemplifying your knowledge of the course structure and its requirements through personal experience. Again there are no rigorous guidelines on how to do this and it is very much down to your own writing style. Whereas some prefer a strict structure, others go for a more synoptic approach, but always remember to be consistent in order to achieve a flowing, easy to read personal statement.

This point ties in closely with writing style. You want one that the tutor will find pleasing to read; and as everyone prefers different styles the only way to assess yours accurately is to show your drafts to as many people as possible. That includes, teachers, parents, friends, siblings, grandparents – the more the better, don't be afraid to show it round!

Despite the lack of a standardised writing method, there is of course a list of standard content to include. In general you are trying to convey your academic, professional and personal suitability for the course to the tutor. This needs to be reiterated whilst demonstrating clear, exemplified knowledge of the course structure and its demands. The biggest problem then in achieving these goals, with all the other candidates also trying to convey the same information, is in producing an original personal statement and remaining unique.

The easiest way to overcome this is to integrate your own personal experiences, reflections and emotions – both demonstrating passion and insight. More practically, it is a good idea to split the main body into two or three paragraphs, in order to avoid writing one big giant boring monologue.

Part One: This should cover why you are suited for dentistry. This will include your main academic interests, future ambitions and what makes dentistry right for you. It is a good idea for you to read up the course syllabuses, and find something that catches your interest above others. If you have read anything outside of the A-level/IB syllabus related to your chosen course which has inspired you then this is the place to mention it.

Part Two: This section should still be about why you're suited to your chosen course with a particular focus on work experience. If you've had to overcome any significant challenges in life and wish to include these in your personal statement, then this is normally the best place to do so. Similarly, any relevant prizes & competitions should be included here. However, it is important to remember not to simply list things. Ensure that you follow through by describing in detail what you have learned from any experiences mentioned.

Part Three: This is the smallest part of the main body and is all about extra-curricular activities. Use this section to demonstrate your manual dexterity skills and other hobbies relevant to a profession in dentistry. It is easy to get carried away in this section and make outrageous claims e.g. claim to be a mountain climber if all you have ever climbed is a hill at the end your street etc. Lying is not worth the risk given that your interviewer may share the same hobby that you claim to be an expert in!

Avoid making empty statements by backing things up with facts. For example: *'I enjoy reading, playing sports and watching TV'*, is a poor sentence and tells the reader nothing. The applicant enjoys reading, so what? Which sports? Doesn't everyone like watching TV? If the applicant is in a sports team, or plays a particular sport recreationally with friends then they should name the sport and describe what their role is. Likewise, the applicant should actually describe how their hobbies relate to them as a person and ideally their subject.

What to Include

Still a little stumped? Here is a summary of a few useful pointers to get you started:

➢ **Sports and other hobbies** – these are particularly important for the vocational courses like medicine and dentistry as they offer a form of stress relief amidst a course of intense studying whilst also demonstrating a degree of life experience and well roundedness. By all means discuss international honours, notable publications or even recent stage productions. Remember to reflect on these experiences offering explanations of how they have changed your attitude towards life or how they required particular dedication and commitment.

➢ **Musical instruments** – Again an excellent form of stress relief and a great example of manual dexterity. Do not be afraid to mention your favourite musical works for that personal touch but also any grades you have obtained thus demonstrating commitment and a mature attitude that can be transferred to any field of study.

➢ **Work experience(s)** – Don't bother wasting characters by citing references or contacts from your work experience but rather discuss situations that you were presented with. Describe any situations where you showed particular maturity/professionalism and explain what you learnt from that experience. It is always advisable to discuss how your work experience affected your view of the subject field, either reinforcing or deterring you from your choice.

➢ **Personal interests within the field of study** – This is a really good opportunity to show off your own genuine interest within the subject field. Try to mention a recent article or paper, one that isn't too contentious but is still not that well known to show depth of reading.

➢ **Personal attributes** – exemplify these through your own personal experiences and opinions. As mentioned previously many courses will list "desired" personal attributes in their prospectus - you must include these as a minimum in your personal statement. Try to add others of your own choice that you think are relevant to the subject in order to achieve originality.

➢ **Awards** – be they national or just departmental school awards, it always worth trying to mention any awards you have received since about the age of 15/16. A brief description of what they entailed and what you learnt from the experience can add a valuable few lines to your personal statement. Providing proof of long term dedication and prowess.

Together, discussion of all these points can demonstrate reasoned consideration for the course you have applied for. This is particularly appealing for a tutor to read as it shows a higher level of thinking by giving your own reflection on the course requirements.

The Conclusion

The conclusion of your personal statement should be more about leaving a good final impression than conferring any actual information. If you have something useful to say about your interest and desire to study dentistry, you shouldn't be waiting until the very end to say it!

Admissions tutors will read hundreds of personal statements every year, and after about the fifth one all start looking very much the same. You should try to make your statement different so it stands out amongst the rest. As the conclusion is the last thing the admissions tutor will read, it can leave a lasting influence (good or bad!) The purpose of a conclusion is to tie up the entire statement in two or three sentences.

A good conclusion should not include any new information, as this should be in the main body. However, you also need to avoid repeating what you have said earlier in your personal statement. This would be both a waste of characters and frustrating for the tutor. Instead it is better to put into context what you have already written and therefore make an effort to keep your conclusion relatively short – no more than 4 lines.

The conclusion is a good opportunity to draw on all the themes you have introduced throughout your personal statement to form a final overall character image to leave the tutor with. Unless there is anything especially extraordinary or outrageous in the main body of your personal statement; the tutor is likely to remember you by your introduction and conclusion. The conclusion therefore is a good place to leave an inspiring final sentence for the tutor.

Some students will make a mention in here about their career plans, picking up on something they have observed in work experience or have encountered during reading. This can be a good strategy as it shows you're using your current knowledge to guide your future aspirations. If you do this, try to do so with an open mind, suggesting areas of interest but being careful not to imply you are less interested in others.

You have to spend a long time at dental school doing general dentistry before you have the option to specialise in any one field, so admissions tutors need to be certain your interest extends into all areas. Secondly, don't sound too fixed about your plans. There is a lot more to see before you can make an informed career choice, so by all means show your particular interests but avoid sounding as though you are closing any options off.

It is important to avoid sounding too arrogant here and over selling yourself. Instead adopt a phrase looking forward in time – perhaps expressing your excitement and enthusiasm in meeting the demands of your course requirements, or looking even further ahead, the demands of your career. For example, consider a phrase like: *'driven by my love of dentistry, I am sure that I will be a successful dentist and take full advantage of all opportunities should this application be successful'* rather than *'I think I should be accepted because I am very enthusiastic and will work hard'*. The sentiment behind both of these statements is positive, however the second sounds juvenile whereas the first is aspirational, confident and yet humble.

Work Experience

Work experience is a great way to demonstrate your commitment to dentistry. It cannot be over-stressed just how vital it is. It is so essential that I have *never* heard a case of someone getting an offer for dentistry without work experience.

During the work experience itself it is wise to keep a notebook or a diary with a brief description of each day, particularly noting down what events happened and importantly what you learnt from them. Whilst there is a designated section of the UCAS form for work experience details, the personal statement itself must be used to not only describe your experiences but also reflect on them. Making sure to discuss the following points:

➢ How did certain situations affect you personally?
➢ How did the experience alter your perspective on the subject field?
➢ Were there particular occasions where you fulfilled any of the extracurricular requirements listed within prospectuses?
➢ Most importantly how did your experience(s) confirm your desire to pursue the field of study into higher education?

Why Work Experience?

Dental schools value work experience so highly because it shows you have a number of essential traits.

Work experience shows you're informed. You're making a potentially lifelong commitment to a profession – so how do you know you'll be suited to it? Rather than basing your ideas of a dental career on the media or stories you hear from others, the best way to convince the admissions tutor you know what the job actually entails is to go and experience it for yourself. Getting as much varied work experience as possible opens your eyes to the work that dentists and other caring professionals actually do, demonstrating you have a realistic understanding of the profession better than any words can. If you have good work experience, admissions tutors are confident you're choosing dentistry for the right reasons.

Work experience shows you're committed. Arranging work experience can be hard – you may need to approach multiple people and organisations before you get a 'yes'. Therefore, if you have a good portfolio of work experience, it shows you have been proactive. It shows you have gone to the effort for the sole purpose of spending your free time in a caring environment. This shows drive and commitment – impressive qualities that will help you gain that valuable place!

Arranging Work Experience

Arranging work experience can be hard. If you're finding it difficult to get exactly what you want, please don't be disheartened. You are facing the same difficulty that tens of thousands of students before you have also faced.

With work experience, it's a very good idea to start early. The earlier you approach people, the more likely you are to be accepted. It is not really practical to start seeking work experience until after you turn 16 due to age restrictions within the work place – especially where confidential information is concerned! So conduct your work experience during the summer after your GCSE examinations and throughout your AS year. This can be achieved through private arrangements you yourself make but it always worth consulting your schools careers officer as well. Remember that any part-time/summer paid jobs also count as work experience and definitely worth mentioning as they show an additional degree of maturity and professionalism.

If you are able to keep up a small regular commitment over a period of months it really helps to show dedication. It's a good idea to always carry a notebook when you're on work experience. Use it to note down anything interesting you see or hear about to make certain you don't forget!

Types of Work Experience

Dental Shadowing

This is where you spend time watching dentists or other healthcare professionals going about their work. It is a great opportunity to learn about the day-to-day activities of people working in dental care. You will see both the interesting and stressful sides of the job. But it can be hard to organise. Aim to get some direct work experience with dentists in practice, or specialists in a hospital or dental hospital setting. If you aren't able to get much experience then you can supplement with any of the other good types of work experience as below.

Dental Volunteering

This includes voluntary work in any dental or care-related setting such as a hospital, dental practice, hospice or care home. Care homes can present good opportunities to volunteer with helping with entertainment activities for the residents, so this is something to consider getting involved in. Alternatively, consider volunteering at the hospital as a "show and tell" guide to help patients find where they are looking for. It doesn't matter that you are not actually providing the care – what matters is that you are in an environment where you are interacting with people who need help and are learning about how they are cared for. Although most seek a voluntary placement, this is not essential; you could take a job in a caring setting. For example, you could work as a hospital porter or a healthcare assistant, either of which would provide a great insight into the workings of the hospital especially if this was in the Oral Surgery or Maxillofacial departments. This can be a particularly good option if you're planning on a gap year: you can work for a few months to gain valuable experience and save up some money, then use the money if you intend on any travelling.

Non-Dental Volunteering

This might include working in a charity shop, library or similar. This is a good way of showing commitment and public service, however, you won't learn about healthcare this way. It can make up the majority of the work experience, but in addition, you need to do something which gives you the opportunity to actually learn about dental care provision.

Becoming a Member

It's good to get as involved as possible with your local health community. By becoming a member of any local health-related organisations, you can demonstrate a willingness to do something to help people. You could become an associate at the hospital, allowing you to vote in Governors' elections. Another possibility is to train as a first aider and join St John's Ambulance (but if you're considering this make sure to plan ahead as sometimes it can take a while to get started). Keep on the lookout for any local health-related projects – councils will often have oral health initiatives such as raising awareness of oral cancer.

Attending Courses

Whilst this is not strictly work experience, this section is the best place to discuss courses. It is not necessary to attend dentistry preparation courses, but there are some good options available. Make sure you check what the course offers before enrolling, and only attend if there is something you will directly benefit from.

Extra-Curricular

It is important to show you are a balanced person, not someone whose only focus is work. Extra-curricular activities can really strengthen your personal statement by showcasing your skills especially manual dexterity. Remember that there is no *intrinsic* value in playing county level rugby or having a diploma in acting – you will not win a place on excellence in these fields. The value comes from the skills your activities teach you. Regardless of whether you're outstanding at what you do or you just do it casually, remember to reflect on what you've gained from doing it. There will always be something positive to say and it may be more valuable than you think.

There are three very important ways that extra-curricular activities can strengthen your application, so make sure to use them to their full capability.

You should use your extra-curricular activities to highlight skills that will help you in dentistry. You play football – talk about how this has helped your teamwork; you draw and paint – surely this has improved your fine motor skills. By linking what you do to the skills you've developed, you take a great opportunity to show the admissions tutor just how well-rounded and competent you are at using your hands skilfully. By showing how you have developed these critical skills you demonstrate, you are a strong applicant.

Interests outside work give you a way to relax. Dental studies and work as a dentist can be stressful, and admissions tutors have a duty of care towards students. By accepting someone who knows how to relax, they are ensuring you'll strike the right balance between studies and relaxation, keeping yourself fresh and healthy through difficult times.

Showing you have enough time for extra-curricular activities can support your academic capabilities. If you are the member of an orchestra, a sports team and you keep a rock collection, you were clearly not pushed to the absolute limit to get the top grades you achieved. For a student without other interests, it might suggest to the admissions tutor they are struggling to keep up with the current workload and may not be able to cope with the additional demands of a dental course.

Skills

Throughout this book we talk about skills – abilities that you have developed through your work and experiences that will equip you to be a good dental student and dentist. Here follows a discussion of some of the major skills it is useful to demonstrate in your personal statement. Remember that this isn't a tick-box exercise – aside from manual dexterity you don't need to provide evidence for every single skill we discuss. This is merely to give you some guidance towards the key skills you may want to focus on as you write your personal statement.

Teamwork

Dentists work as part of a large team comprising of different healthcare professionals such as nurses, dental therapists, hygienists and technicians. In order to achieve good outcomes for the patients, effective teamwork is essential. You can demonstrate your teamwork through team sports, musical ensembles and collaborative projects.

Manual dexterity

Dentists are constantly performing demanding treatments in a limited space with a high level of skill. You can demonstrate or develop the required excellent hand eye coordination and fine motor control by playing a musical instrument or taking part in arts and crafts activities such as drawing or jewellery making.

Time-management

Dentists are very busy. As you may have seen in your work experience, they constantly have a long and varied task list including prescribing, patient assessments, referrals, treatments, phone calls, meetings and more. As such, they need to be able to manage their time well to prioritise the most important tasks, and to work quickly and efficiently to make sure they complete everything that needs doing. Show your time-management skill by balancing a number of different activities at the same time.

Working under pressure

Dentists can work under a lot of pressure, whether this is because they have so much to do or because some particular task is critical to get right. It's important that dentists can remain calm under pressure so they work well and avoid becoming too stressed. Show that you can work under pressure by working as a first aider, competing in high-level sport, becoming a sports official, taking responsibility for something important in a group activity or helping run an event with a committee or society.

Communication

Good communication is one of the most important skills that a good dentist has. This is essential both in talking to patients and the rest of the healthcare team. Talking to patients, it is important to be able to explain clearly what is going on and to be able to answer their questions. Dentists also need to be able to put patients at ease, and manage very anxious or phobic patients. Talking to the rest of the healthcare team, it is important to be able to communicate plans clearly and suggest your ideas in a logical way. You can show your communication proficiency in a vast number of ways – absolutely anything where you need to talk to communicate ideas to another person. If it is under any particular pressure then bonus points!

Self-directed learning

Some dental schools have a very structured teaching programme. Others, such as the PBL (problem-based learning) courses rely much more on individual research and learning. But at any dental school and then afterwards as a dentist, you will be required to use your own initiative to both work out what you need to learn and then to learn it. Therefore, you need to show you have this ability to seek out and learn information for yourself. Demonstrate this ability by reading outside the normal curriculum, by reading books and articles about dentistry or by taking on an extra self-taught course or module.

Organisation

Similar to time management, but dentists have to be organised to ensure they never forget about something important to do. Show your organisational skill by being a member of a society or committee, planning some form of event, organising a group activity or a teaching scheme.

Leadership

Dentists have to be leaders of the dental team, making treatment decisions and discussing plans with others involved in the care. Therefore to make a good dentist, it helps to have some natural leadership ability. Show your leadership skills by taking a prefect role in school (describe what you're responsible for, don't just name-drop the title), by leading a project, by sitting on a committee, by conducting a music group or by captaining a sports team. DoE (Duke of Edinburgh) awards can be a good way of showing leadership as well as a number of other positive personal qualities.

Teaching

Training as a dentist is a long process, and even once qualified, dentists embark on a process of lifelong learning to advance their knowledge and keep up to date with new developments. All through the training process, more senior dentists help by teaching dental students and supporting foundation dentists. Therefore, teaching is an important skill of any dentist. Show your teaching skill by presenting a project in school, helping to teach younger students, teaching a musical instrument or helping with sports coaching, for example.

Achievements and awards

Not specific skills of course, but dental schools are looking for the very best applicants. If you have won any particular prizes or awards either through school or though a society, or you have done particularly well in the UKMC (Maths Challenge) or a science Olympiad, then it would be great to mention it on your personal statement. It's just one extra thing to help your abilities stand out from the rest.

DEFERRED ENTRY AND GAP YEARS

It is always advisable to apply to university during your A2 year – at the very least it is a useful experience and you can always apply again next year if you are unsuccessful. In attending university a year later, you are a year older, bringing more maturity and life experience to the course – the benefits of this are clear to see in course like dentistry!

If you are planning to take a gap year, always apply whilst in A-level year unless there is a reason you would not be able to gain a place (e.g. grades/predictions too low, you need to sit more exams). Applying for deferred entry allows you to go on your gap year, safe in the knowledge that you have secured a place upon your return. If things then don't go to plan, you have time to improve your application and a second chance in which to apply to different dental schools.

You'll need to tweak your statement slightly if you're applying for deferred entry. You will need to demonstrate to the tutor that you are filling your gap year with meaningful experiences in order to help you grow as a person. Therefore discuss your gap year plans in a brief paragraph, describing what you hope to achieve, what life skills you hope to learn, and how these are both transferable and applicable to your course. In addition, a year of deferred entry gives you opportunity to work and save in order to fund your progress through what is a very expensive time at university.

This is a good opportunity once again to show your commitment. If your gap year plans include any volunteering work, use this to support your vocation of public service. If you have already made plans, it shows that you're organised.

To make a strong application, you should be spending a significant proportion of your gap year doing things that support your application: work experience, voluntary work and activities that build your skills. Discuss in your personal statement why you chose to do these things, what you are learning from them and how it has affected your desire to study dentistry. Make sure you account for all time and give reasons for everything you do, tying it back to your path towards (hopefully) becoming a dentist. A good application should draw upon your gap year to reinforce your skills and commitment; it should give positive reasons why you have chosen to take a gap year. Taking a gap year gives you good opportunities to expand your experiences, but you have to remember that it also brings expectations – therefore if you don't take these opportunities you stand to weaken your application.

Going on a gap year is a choice for you to make; overall, you are equally likely to get an offer with or without taking a gap year.

Re-applying

If instead you are reading this during a gap year because of an unsuccessful first application do not be disheartened. Applying a second time puts you in a much stronger position as you have your A-level grades in hand. Do mention your failure first time round in your personal statement, but also reflect on it and discuss why you think this happened. More importantly, discuss what you have done to address these issues to improve yourself as a candidate. Re-applying shows strength of character, resilience and determination- qualities desired by any course tutor at any institution.

MATURE AND GRADUATE APPLICANTS

If you're applying as a mature student or to graduate entry dentistry, talking about your previous work and career is important.

If you have been working for a number of years, then a large chunk of your relevant life story will be due to your employment. Your journey to dentistry will describe your previous career path, the moment you thought about a change, how you investigated dentistry as a career and why you now believe it is the right path for you.

Coming from a professional background, the skills you have learned in the workplace will be significant and begin to overtake extra-curricular activities as a way of demonstrating core attributes such as time-management, communication and team working. In addition, you may have undertaken professional learning in your job such as reading books or attending courses – be sure to draw upon this to support your ability to undertake the lifelong learning required of a dentist.

Admissions tutors are not looking to see how similar your current job is to dentistry. You will learn what you need to know at dental school. They are looking for the general skills you have learned that will help make you a good dentist, and research/experience outside of work to confirm your interest. There is, however, an exception to this – if you work within science and are applying to 4-year accelerated post grad dental courses, be sure to talk about your scientific education. Describe the scientific skills you have learned such as data interpretation, because this shows your aptitude for science and your suitability for picking up new technical knowledge quickly.

It would also be prudent to contact the admission tutors to check the requirements for entry if applying as a mature or graduate as it will differ dependant on the university. As with undergraduate entry each university will have their own strict criteria and therefore making sure they will consider your application first will save you from wasting your precious application slots when applying through UCAS.

Extra Reading

Reading above and beyond what you would need to for your school studies is a great way to show genuine enthusiasm. Therefore, a good personal statement will include at least some discussion of this extra reading.

Make sure you don't fall into the trap of thinking a long list of books will impress – this isn't the point. **The idea is you show what you have learned** from each of the books and how it has influenced your decision to study dentistry. This shows that you haven't just looked at the pages of the book as you've turned them over, but rather that you have understood and thought about them. When discussing your learning, try to make specific points rather than generic ones. For example, a weaker statement might say:

"I read Thinking Fast and Slow by Daniel Kahneman, which helped me understand the way decisions are made".

Whereas a stronger statement may say:

"I particularly enjoyed Thinking Fast and Slow by Daniel Kahneman, which made me realise the importance of shortcuts in making quick and accurate decisions."

Standing out from the crowd

You may have heard people saying that a good personal statement helps you stand out from the crowd – and this is certainly true. Admissions tutors read hundreds of personal statements, so to be in with the best chance yours should offer something a bit different to leave a lasting good impression.

Whilst standing out from the crowd is easy, the line between standing out for the right and the wrong reasons is a fine one and you have to tread carefully.

The easiest ways to add some originality are in your reading and activities. There will be countless people who play football but less who play ice hockey; everyone reads New Scientist and Student BDJ (both excellent resources you should look at) but fewer people read Nature. It is not more valuable to do something less popular, but it can make it easier for the reader to see your personal statement as original. This is not about going out and enrolling with an extreme ironing club – it is about taking time to identify the things you already do and skills you have that are a bit more interesting than the generic activities and just giving them a mention to show a wide variety of interests.

Many dental schools will score the personal statement based on a marking grid. You'll gain marks for evidence of performance in different areas depending on your assessed level of achievement. These areas may include interest in dentistry, variety of work experience, evidence of altruism/volunteering, communication skills and general skills. Ensure you cover all the areas described in the section guide to make sure you hit the entire key scoring points.

Proof-reading the personal statement is extremely important – not just you, but also by showing it to friends, family and teachers to get their opinions. Firstly, it's so easy to ignore your own mistakes, because as you become familiar with your own work you begin skimming through rather than reading in-depth. But also, this allows people to assess the writing style – by gathering lots of opinions you can build up a good idea of the strongest areas (which you should expand) and the weakest areas (which you should modify).

Don't try to force anything into the personal statement. Allow it to grow and showcase your wide variety of skills. Make sure there is a smooth flow from one idea to the next. Allow it to tell your story. Make sure all the spelling and grammar are accurate. Then, your personal statement will shine out from the average ones to give you the best possible chance.

Omissions

It can be difficult to work out exactly where the line stands when it comes to omitting certain information - sometimes leaving certain things out can cause problems.

For example, let's imagine you worked for half an hour a month at a care home over a 6 month period. If you said in your personal statement you had worked at a care home for 6 months, you could reasonably expect interview questions on it. If it emerged that you had only spent three hours there in total, the interviewer would be left doubting the truthfulness of the whole personal statement.

Another circumstance when not to omit details is when there is something that needs explaining. Perhaps you've taken a year out of the normal education pathway to do something different or because you were experiencing some difficulty. Whilst the personal statement is not the place to discuss extenuating circumstances, it should tell the story of your recent path through life. If there are any big gaps, it is likely to concern the person reading it that you have something to hide. Make sure you explain your route and the reasons for it, putting it in the context of your journey towards a dental career.

Interviews

As a dental applicant, you are almost certain to be interviewed before any offer is made – and this fact adds extra complexity to the writing of your personal statement.

In any interview, you can expect to be asked questions on the content of your personal statement – about your work experience, your reading, your extra-curricular activities and so on. This makes it especially important to be completely honest. And I don't mean just avoiding *explicit* lies (you shouldn't be doing that anyway; dentists and dental students are legally bound by a moral code) – this includes all the little traps that are so easy to fall into – the book you intend to read, the operation you 'watched' but didn't fully see and so on. That book you were genuinely *planning* to read might turn out to be terrible, but you're then committed to reading it front to back in case your interviewer probes your interest in it. Likewise, if you couldn't see an operation in your work experience because there were 5 broad-shouldered surgeons blocking your view, it might be sensible to avoid bringing it up. Your interviewer might be a specialist maxillofacial surgeon and ask all about maxillomandibular advancement, leaving you feeling silly if you cannot answer about an operation you claim to have watched.

But this isn't all bad news – **it can actually be a very positive thing**. By writing about all the subjects that interest you most in your personal statement, you have the opportunity to guide the interview discussion towards those areas you love, know most about and would enjoy discussing. By doing so, you give yourself an opportunity to show your knowledge and enthusiasm to the interviewer – traits which will go a long way in convincing them you are the right person to fill that elusive place at dental school.

Therefore, it is important that you use your personal statement as part of your interview preparation. Read and re-read your statement before the interview to make certain you are ready to talk about anything you may be asked questions on. Not only does this give you a great chance of answering these questions well, it can give you an overall feeling of assurance that you are well prepared, lending confidence to make your overall performance more polished. What's more, if you have all your personal statement information at the front of your mind, answering general questions about your experiences is much easier as you have a great bank of information to quickly draw upon.

Things to Avoid

Whilst there are no rights and wrongs to writing a personal statement, there are a few common traps students can easily fall into. Here follows a discussion of things that are best avoided to ensure your personal statement is strong.

Stating the obvious – this includes phrases like "I am studying A-level biology which has helped me learn about human biology in the human biology module". Admissions tutors can see from the UCAS form what A-level subjects you are studying and the learning you claim is obvious.

University names – the same personal statement goes to all universities, so don't include any university names. Only include specifics of the course if they are common to all the courses you are applying to.

Controversy – avoid controversy in any form, be it strong opinions or any other reason. You don't want to make an impression for the wrong reasons, and if you irritate the reader you're making life needlessly difficult.

Lists – everything needs to be included for a reason. Very few things have an intrinsic value, rather the value comes from the knowledge you gain and the skills you develop by doing the activities. Therefore, reeling off a long list of sports you play won't impress anyone. Instead, focus on specifics and indicate what you have learned from doing each thing you mention.

Detail about your A-level subjects – most dental applicants study similar A-levels and they are included on your UCAS form. This won't help you stand out and is a waste of words.

Things that happened before GCSE – if something *started* when you were nine and you have continued it up until today then you should absolutely include it as it shows great commitment and the opportunity to develop many skills. However, if you are considering mentioning the archery you stopped four years ago, please resist the temptation. Putting something that finished a long time ago signals to the reader that you don't have much going on now – not the impression you want to be making.

Include books you haven't read – this is risky. Even if you *genuinely* intend to read the book, you can't make any intelligent observations about it if you haven't done so yet. In addition, you are then committed to finishing it even if you find it very dull, or you risk being caught out in an interview. Stick to things you have already read. If you don't have much to say, pick some short books and journal articles and make a start today!

Extenuating circumstances – the personal statement is to tell your story. It is not the place for extenuating circumstances. If any are applicable, this is for teachers to write in the reference. Make sure you know who is writing it and meet with them to help explain the full story.

Plagiarism – it goes without saying that you must not plagiarise, but I feel no "things to avoid" list would be complete without the most important point. Plagiarism of another personal statement is the easiest way to get yourself into big trouble. UCAS use sophisticated detection software and if any significant match shows up (not necessarily the whole statement, just a few identical sentences is enough), then universities you apply to will be notified and are likely to blacklist your application.

The Reference

The UCAS reference is often neglected by many applicants; it's an untapped resource that can give you an edge over other applicants. In order to plan your use of the reference you first need to establish how it will be used – again consult prospectuses or subject websites. Does it actually count towards your application score or rather is it only consulted in borderline candidates? Furthermore the reference could certainly affect the way in which the tutor perceives what you have written and indeed what they infer from it.

Either way, in order to get the most out of your reference you need to actively participate in its creation. The best way to achieve this is to ask a teacher whom you are particularly friendly with to write it. Even if this is not possible, ask for a copy of your reference before it is submitted to UCAS. This way you can ensure that the personal statement and reference complement one another for maximum impact.

The reference is best used for explanations of negative aspects within your application – e.g. deflated exam results, family bereavements – or even addition of new information if you run out of space in your personal statement. In this respect the reference is a backdoor through which you can feed more information to the tutor in order to strengthen your application.

If there is a teacher who is willing to go through your reference with you, complete your personal statement first before starting on the reference itself. This way you will have a clear idea of the content and tone of the majority of your application as well as anything that may be missing which you would like to add.

The reference is the one place for your teachers to be completely unreserved- superlatives and complements mean a lot more coming from someone other than yourself. One such example of this is the opportunity for your teachers to discuss how they have actively noticed your initiative and passion, going above and beyond in pursuing the subject in question.

EXAMPLE PERSONAL STATEMENTS

Below you will find five example personal statements, with comments drawing your attention to the stronger and weaker points of that personal statement.

To make best use, don't look immediately at this. First, read the personal statement yourself and get a feeling for the general style of writing. Then, test yourself: decide which you think the strongest and weakest parts are. After that, look at the comments on the statement. By using the book this way, you develop your own critical reading skills – skills which you can then apply to your own personal statement, allowing you to build in improvements.

IT CANNOT BE OVERSTRESSED HOW IMPORTANT IT IS THAT YOU
DO NOT COPY FROM THESE PERSONAL STATEMENTS

Rather use them for inspiration. Plagiarism is a breach of academic responsibility and is highly likely to be detected by UCAS's *Copycatch* software.

■ ■

Statement 1

Dentistry offers me the opportunity of a challenging career that will develop my interest in science, while helping others. After receiving orthodontic treatment I became intrigued by the capabilities and results of dental procedures. I am ready to apply myself to hard work to achieve my ambition, working as part of a dental team serving the public.

My work experience in both a general and specialist dental practice reinforced my view that a career in dentistry would best use my skills and interests. I also observed that teamwork and interaction with work colleagues, is just as important as the dentist patient relationship. My time there gave me an insight into various treatments, for example fillings, extractions and crown preparation. I was allowed to take impressions of model teeth so I could prepare the tooth for a crown and make a temporary crown. I realised how technically demanding and intricate, dental procedures can be. At the specialist orthodontic practice I viewed the provision of a variety of appliances from fixed to removable ones. I also gained an insight into the management side of this profession and the continual changes that occur within NHS contracts.

My interest in science has been heightened studying both chemistry and biology at AS level. In chemistry recently we synthesised Aspirin, which I found fascinating. We verified the purity of the final product by determining the melting point, then comparing it with the actual value for pure Aspirin. In biology while learning about the anatomy of the heart, I found the dissection of the lungs and heart of a lamb intriguing. Later we studied causes of infectious diseases, and many were preventable, which I was especially interested by. I have read articles, which show that poor dental hygiene and gum disease may contribute to cardio vascular disease and premature death. I find it appealing that dentists administer a whole cycle of care, with initial examinations, treatments and reviews.

I enjoy working as part of a team and interacting with the public. During the summer I volunteered for a week at Vitalise, a respite care centre for disabled people. I had to build a rapport to gain their confidence and vary my methods of communication to suit their needs. I enjoyed building a trusting relationship with them and making sure they had a pleasurable stay. Since 2005 I have volunteered at my local Brownie pack. The group leader believes I have become a very valued member, capable of taking sessions independently, especially creative activities, which the Brownies are also enthusiastic about. Over the past 5 years I have completed my Bronze, Silver and Gold, Duke of Edinburgh award. Teamwork and perseverance were important skills, on our recent expedition when my group member sprained her ankle. We all worked together to help her back to camp safely. To maintain her morale I stayed at the back of the group with her. I learnt essential skills such as first aid and the importance of planning.

Dentistry involves precise work in confined spaces with great skill. I have a passion for art, making distinctive pieces. I work with a range of materials including glass and acrylic paints. A variety of artists inspire me, especially Janet Fish, as I find her style unique. My other interests are playing jazz pieces on the piano and flute.

My ambition is to become a dentist. I believe all my attributes are well suited to this career. The responsibilities and skills I have gained from Brownies, Vitalise and the Duke of Edinburgh award awards have prepared me to become a well-rounded person, capable of achieving my goal. As a focused and hard working individual I am enthusiastic about contributing to university life and look forward to pursuing dentistry as a career.

Universities Applied to:
➢ Sheffield: Offer
➢ Leeds: Offer
➢ Birmingham: Offer
➢ Newcastle: Rejected

Good Points:
Well written and structured. There is good insight gained from the work experience that has been reflected on sufficiently. Additionally the students' hobbies have been clearly linked to their relevance to dentistry demonstrating the desired personal qualities of a dentist. The first and last paragraphs also provide a good reason of why dentistry was chosen and why the student would be a great fit to pursue a career as a dentist. Although there is a large portion devoted to discussing the students A level studies this has been nicely tied to a very topical dental subject – the relation between oral health and cardiovascular disease.

Bad Points:
The opening statement is quite generic and does not hold the attention of the reader, the students own experience of dentistry could also be developed. There are parts where the statement does not flow as the hobbies are discussed across multiple paragraphs.

Overall:
The statement is strong as the student has managed to tie her experiences to her motivation for studying dentistry and clearly explained why she is suited to this career path. A more punchy opening statement would have grasped the reader's attention from the start.

Statement 2

I became interested in dentistry when my grandmother was diagnosed with mouth cancer. Unaware that a routine check-up can lead to major diagnosis I became intrigued. After extensive work experience and investigation I have decided that dentistry alone provides the best combination of my existing interests in science, ability to interact with patients and utilise surgical skills at the same time. As a dentist I will not only have opportunities to reassure and educate patients with complaints like toothaches but also diagnose serious ailments such as oropharyngeal cancer, thus improving the quality of my patients' lives.

I undertook nine months work experience at NHS and private practice to gain further insight and was elated to attend a week at orthodontics and cosmetics. I learnt dentists not only interact with patients but also work in a multi-disciplinary team with nurses, technicians and orthodontists. Such teamwork ensures efficient healthcare and means workload is divided appropriately. I enjoyed observing treatments such as crowns, fillings and extractions where I realised that a dentist's confident and reassuring demeanour puts patients at ease. Observing an implant placement was an exciting experience which brought to my attention that fine motor skills are crucial in the accuracy of such procedures. These insightful experiences and enjoyable patient-contact confirmed my decision to become a dentist.

In Chemistry titrations taught me patience, a vital skill needed to perform long procedures. This also developed my time management skills and an ability to work under pressure. Studying the diversity of human behaviour in Psychology will help me, as a dentist, to flexibly deal with patients of different temperaments and cultures. In Biology, I learnt the anatomy and physiology of human body and the adverse effects of smoking and alcohol. Although dentistry primarily focuses on oral cavity, I believe a good dentist has an awareness of other systems in the body and are able to link dental complaints to other diseases. In English, I chose to be a team leader during several projects where I demonstrated my ability to lead a team by exercising decision-making and delegation of roles, vital because as a dentist I will be guiding my team effectively for the ultimate care of patients.

Currently I am voluntarily working in a team at North Manchester Hospital to improve functional mobility of stroke patients. Although challenging and a lot of responsibility, it is also very rewarding to see patients improve coordination after repeated exercises. Taking a First Aid course with St John Ambulance provided me with essential knowledge of handling casualties. Spending two months with the elderly at a Day Care Centre improved my interpersonal skills and working at Oxfam allowed me to support my community.

As a Bury Youth Cabinet member I express my views over changes in Bury. I recently raised funds for Somalia at my local Mosque where I managed a team and organised tasks ensuring successful outcomes. There is no doubt that dentistry is a stressful profession however, with my temperament and hobbies of swimming, playing badminton and reading books I will definitely be able to cope successfully. I took an opportunity to attend dissection club where I applied observational skills and perfected my motor skills using various instruments to dissect mammals. My hobbies of sewing and cushion making have also developed my manual dexterity which is a useful tool for a skilful performance needed in dentistry. Moreover, I successfully made a crown at a dental laboratory and realised that immense concentration and control are crucial for dentists to perform procedures in confines of the mouth, where precision is essential.

Universities Applied to:

- ➢ Sheffield: Rejection
- ➢ Kings: Offer
- ➢ Birmingham: Offer
- ➢ Liverpool: Interview + Rejection

Good points:

Well-written which allows the statement to flow nicely. A very personal entry from the start immediately grabs the reader's attention and makes you want to read on. Giving insight into what motivates the student to pursue a career in dentistry is centrally important. Having a concrete case to tie this motivation to is helpful, as it gives the statement a human and individual touch and also provides material to discuss during the interview. The student also displays extensive work experience, which is a strength. Linking her experiences back to dentistry shows she is aware of the challenges and skills needed in this career. The applicant also shows various hobbies and tasks to demonstrate manual dexterity.

Bad points:

While having a good introduction there isn't an obvious conclusion to the statement. While not a major problem it would really round off and summarise why this applicant would be suitable for a career in dentistry and reinforce the reason they wish to pursue it.

Overall:

Overall this is an excellent statement. The strengths definitely lie in the personal touch with the motivation to study dentistry. The student links her experiences that are relevant to studying dentistry showing not only does she understand the rigors of being a dentist but also that she has the skills to be able to succeed.

NOTES

Statement 3

Dentistry provides a jigsaw of intellect, practical experience and endless opportunities. Enthralled by the infinite techniques used to transform my mouth through orthodontic treatment, I became passionate about studying dentistry. Having the confidence to smile again instilled in me a desire to create this confidence in others. Researching in detail the effects of halitosis for my EPQ has broadened my knowledge in this field, whilst studying Chemistry and Biology have highlighted the necessity to organise and execute investigations with precision and accuracy.

Eager to experience the dentistry field, I shadowed three GDP'S and dental hygienists. Watching the dentist direct the hygienist, conduct procedures and provide the patient with a comfortable atmosphere emphasised the importance of being able to control numerous factors and highlighted that teamwork is essential in ensuring better oral healthcare. Through hosting Challenge of Management I have developed my ability to simultaneously handle a number of issues as well as working effectively as a team. Observing simpler tasks such as denture impressions and extractions was interesting, yet technically demanding tasks such as a root canal filling particularly intrigued me, as I was surprised with the range and precision required. Additionally, when faced with patients expressing emotions of a varying nature, the dentist adapted to suit the situation at hand. I found this quality compelling and sought to demonstrate my organisational, leadership and adaptability skills by completing a Sports Leadership Award where I led, planned and assessed. When in surgery, it was evident that the numbers of patients seen were far more than anticipated, helping me gain valuable insight into how stressful and challenging a dentist's day can be. I know if presented with a challenge, my willingness to succeed would lead me to find a solution.

Over six months I volunteered at a centre for deprived children and a Hospice. Through compassion and perseverance I watched how the smallest things, such as caring, could make the greatest difference. I liaised with diverse members of the public and staff, often in Urdu, strengthened my team work and effectively handled challenging situations and also enhanced my communication skills as well as bonding and gaining the trust of the children; a task difficult to achieve. At school, the role of a prefect has honed my capability to guide and implement rules and responsibility through effective interpersonal skills. I also volunteered with a local community programme dedicated to children with disabilities, where I learnt to express creativity and improve my social and adaptability skills. Furthermore, I received an award for my time spent mentoring students.

My passion, enthusiasm and commitment for designing secured my place from four in 150 students to compete with designs of older European students in Belgium. Involving ICT skills, long periods of concentration and intricate drawings requiring great manual dexterity; skills vital to dentistry. In addition, this trip allowed me to experience diverse cultures and create a network of friends, which I also obtained through swimming, a hobby since the age of seven. Fitness for me alleviates stress in a healthy way and after attending Bollywood dance workshops, performing on stage is now a new hobby. Also after completing a Level 2 Award in Food Safety and Hygiene, I often cook and occasionally sew.

Always a keen enthusiast, I embark upon each task clear headed and optimistic. I have a strong desire to pursue a career in dentistry and I have the determination, the zeal and the qualities required to become a successful dentist.

Universities applied to:

➢ Leeds: Offer
➢ Cardiff: Rejected
➢ Bristol: Rejected
➢ Liverpool: Offer

Good points:

The student demonstrates some good reflections on their work experience. This is very relevant as the experience only really becomes relevant for providing strength to the statement if it is put into the right context and met with adequate reflection from the student's side. The student correctly underlines the correlation of soft and academic skills in the practice of dentistry. This is important, as it is a commonly underestimated relationship. In addition to the clinical work experience, the student also provides a good range of non-clinical experiences that all contribute to their personal development. Particularly relevant in this context are lessons learned teaching as well as communication skills.

Bad points:

The very first sentence falls into the category of filler and does not tell you any useful information but is a good example of stating the obvious. In addition, at instances, the statement lacks a clear structure and a clear message. The information provided is a little all over the place. This is a pity as the unorganised structure makes it very difficult to follow the content and learn about the student, which significantly weakens the overall expressive power of the statement. The second paragraph in particular would have been improved by splitting up the work experience from the other content. Use of words like 'occasionally' should be avoided as this may give the wrong impression

Overall:

A good statement that is, unfortunately, let down slightly by some stylistic and content weaknesses that make it difficult to draw the maximum amount of information about the student from the statement. With some more or less minor improvements to structure and depth of reflection, this statement could be very strong.

NOTES

Statement 4

For me dentistry offers an academically and mentally challenging profession which amalgamates my fascination with the oral cavity and my desire to work with a variety of individuals with their own individual problems on a day to day basis. It offers a chance to make a real difference to the lives of others.

Additional reading has fuelled my passion for the subject, namely a book on the brain such as Peters 'The chimp paradox' This has given me a good grounding in understanding the human brain and recognising how the mind works, in order to manage your emotions and thoughts. I have also borrowed past A level textbooks from school that cover areas of the syllabus that have since been cut such as the anatomy and function of the human eye and I keep up to date with dental affairs using the Science and Health sections of the BBC news website.

I have explored my interest in the subject through work experience. My first placement was with an oral medicine consultant who specialised in oral disease. The one-on-one consultations showed me the all-important need for tact when dealing with difficult issues that needed to be addressed. It also highlighted the great potential for progress in dental research which is exciting for me.

During my second week long placement in a general dental practice I observed a GDP dealing with a broad spectrum of individuals who presented cases ranging from acute infections, to problems with dentures to even minor oral surgery procedures such as extractions. For me, it emphasised the range of skills required to be a dentist: the knowledge of the physiological systems that underpin each diagnosis and how specific treatments will help prevent further problems or restore a patients oral health, knowledge that has a constant need to be replenished due to advances in dental research and technology; the vital interpersonal skills and clarity of communication required to convey what may be a complicated concept to someone with little scientific or dental knowledge; and the manual skills involved in thoroughly examining patients and carrying out minor operations and restorative treatments.

I volunteer weekly in a residential care home for severely disabled adults, most of whom have acute cerebral palsy. Brushing the teeth of the residents has taught me a lot about the value of patience and empathy in dealing with the seriously disabled and what a challenging task this can be for some patients. I thoroughly enjoy getting to know the habits of many of the residents and although none of them has any coherent method of communication each individual has a unique personality that I have come to appreciate over time. While challenging, finding unconventional methods of communication with the residents is very rewarding. Working with these people has really made me realise that I want to devote my life to using my intelligence, diligence and enthusiasm for the good of others; I think dentistry is a natural career choice given this perspective and my desire to apply my manual dexterity skills.

I believe my extracurricular activities have taught me valuable skills that will prove useful as a dentist. I am part of the Nottingham Youth Orchestra and the East Midlands Youth String Orchestra. Playing as a part of these ensembles requires fine motor skills as an individual, as well as an ability to coordinate delicately with the many other members of the orchestra. In addition, during my weekly shift at a restaurant I have the role of training new employees which highlights my ability to explain with clarity and to be friendly and welcoming.

Work experience and speaking with dentists and other members of the NHS has made me appreciate how challenging a career in dentistry will prove yet I am certain that this is the right choice for me as it offers personal challenge, continual development and the opportunity to make a real difference in people's lives. I hope you will give me the chance to fulfil this aspiration.

Universities applied to:

> Liverpool: Offer
> Newcastle: Offer
> Birmingham: Offer
> Bristol: Rejected

Good points:

This is an excellent statement. It is well-written and well-rounded providing a wide range of insight into the educational career of the student. It also gives a good impression of previous work experience and the student ties these experiences well into the whole picture of dentistry. It makes it clear how these experiences have contributed to their choice of dentistry as a subject which is very helpful for an examiner reading this statement. The student also ties his past work experience to lessons learned that they see relevant for dentistry. This is important, as work experience can only be useful if it teaches relevant lessons.

Bad points:

The paragraph addressing the interest in the scientific side of dentistry is somewhat superficial. Whilst it is good to show interest in anatomy and a desire to stay up to date with current dental developments, this is also something that is expected from students aiming to study dentistry. It therefore serves little purpose as a distinguishing feature from other applicants.

Overall:

A very good statement that ticks all relevant boxes and only has a few minor weaknesses. These weaknesses have little impact on the overall quality as the student manages to demonstrate a variety of lessons and experiences that support their choice of dentistry as a career.

NOTES

Statement 5

The first time I announced I wanted to be a dentist; my parents were amused but indulgent. Their reactions are understandable, considering that I was eight at the time. From a young age I have always been intrigued by teeth and it has only grown from that time. My fascination with science is one of the reasons I want to study Dentistry. The continuous learning throughout my career; constant new discoveries and technologies; as well as the variety in each day are part of the attraction of Dentistry.

To form a realistic image of a profession in Dentistry I have undergone various work experience which has allowed me patient contact and a chance to observe professionals. I arranged my first two-week placement at St James Hospital, where I learnt basic practice such as, data confidentiality and hand hygiene which is becoming more important with the emergence of the new superbug, NDM-1. I had another two-week placement in a general dental practice, where I gained knowledge of how the management and administration of a practice operates as well as observing a GDP. This is useful knowledge for understanding how a practice is run and the day-to-day work of a dentist. My work experience has strengthened my resolve to pursue a dental career.

My A level choices confirm my enthusiasm for science and demonstrate that I am able to cope with a heavy workload and rise to a challenge, which have already resulted in an achievement of an A* grade in my A-level Mathematics. I enjoy reading and keeping up to date with the latest developments in science; I am a subscriber to "Biological Science Review" and regularly read the "New Scientist". I am currently writing an EPQ on the ethics of organ donation which is self-motivated and gives me a chance to be in charge of my learning. I am also the creator and president of the Dental Debate Society at my school. We meet weekly to discuss common dental controversial topics.

I try to balance my interest in science with a variety of other activities. As a Senior Prefect and a School Council Member, I have excellent organisation, time management and leadership skills, along with the ability to negotiate. My communication and listening skills have developed through Charity Committee, debating and Netball. As a amateur jewellery maker I have refined my fine motor skills and attention to detail. I am a philanthropic individual and enjoy assisting others. I am a volunteer at my Sunday school and local library. Paired tutoring is a scheme I am also involved in, where I help a younger student who has difficulty reading. Taking part in the Duke of Edinburgh scheme has shown me the importance of perseverance and motivation to succeed.

I am a focused and determined person with a fierce commitment to studying Dentistry. I believe I have the academic capability and drive to succeed in a Dentistry course at university. My aspiration is to become a Paediatric dentist and one of the top experts in my field.

Universities applied to:

➢ Liverpool: Offer
➢ Glasgow: Offer
➢ Newcastle: Offer
➢ Queen Mary: Offer

Good points:

A strong, well-written statement that demonstrates a varied history of academic excellence. The student provides good insight into how the early desire to be a dentist has shaped their development, both academically as well as individually. This demonstrates great dedication to the subject matter as well as the intellectual and motivational facilities necessary to perform well in a demanding course such a dentistry. The student demonstrates good academic performance and discipline.

Bad points:

The statement is very focussed on academic performance and academic detail. Personal experiences and lessons learned during patient exposure are somewhat limited, which is a pity as the student shows considerable clinical experience. It would complete and strengthen the picture of academic excellence significantly if the student had been able to add clinical and inter-person lessons learned during their time in the hospital. This includes skills such as communicating information which is essential in dentistry.

Overall:

A good statement providing good insight into the impressive academic performance of the student. Unfortunately the student sells themselves somewhat short by ignoring the non-academic side of dentistry that is equally as important as academia. Having had the hospital exposure, it would have been easy to add this in order to achieve an even better statement.

NOTES

Statement 6

Trying to define yourself is like trying to bite your own teeth'; a tricky challenge. Challenges simply fuel my ambition, as shown by my dedication and passion to my academic studies. I used Alan Watts' quote because of my desire to enter the world of Dentistry, which stems from personal experience at the Dentist's, where my fears were allayed prior to an extraction under local anaesthetic. The Dentist's ability to achieve trust from anxious patients, like myself, was inspirational, and fuelled my drive to reverse the roles and sit in the Dentist's chair.

Ongoing work experience in the NHS and private sectors, has exposed me to a wide variety of treatments and ethics in the profession. I became aware of the dentist's ability to identify and resolve problems via clear communication in addition to having an acute ability to listen. Being afforded the opportunity to make dental moulds, bite blocks and dentures allowed me to utilise my manual dexterity and creativity, whilst using specialist equipment. Shadowing special needs dentists, as well as dentists at emergency dental clinics highlighted the compassion and patience required to assure patients, especially those suffering from sensitive and sometimes awkward illnesses such as mouth and neck cancer. It was inspiring to see how the dentist spoke to a range of patients at their particular level, in addition to his team. This was evident when the dental team had to revive a patient suffering cardiac arrest; an astonishing moment to witness. I thoroughly enjoyed my work experience and look forward to my placement at the city hospital where I will shadow sessions in theatres, wards and clinics.

This will supplement the 185 hours experience I have accrued thus far. Over the last two years I utilised my bilingual skills working with members of the public in a local Pharmacy. Working within the pharmaceutical team has demonstrated the importance of responsibility, dedication and the need for confidentiality. Preparing needle packs for drug users has exposed me to wider issues and has required my utmost attention to detail. Dental controversies of today, such as fluoridation and amalgam fillings drove my Extended Project; 'Is the fluoridation of public water systems ethical and justified? Reading the magazine 'The Dentist', I have discovered intriguing and informative articles relating to Dentistry.

Membership of the Equality & Human Rights Youth Network enables me to organise high profile events and activities, including raising racial awareness in schools, improving my leadership and organisational skills to a large degree. Mentoring Gifted students through their GCSE's, and conducting a discussion group for Gifted year 8 students has helped me command the trust required to work with younger people. These skills were utilised during my work at a refuge for vulnerable and abused women. My caring and empathetic nature stems from nursing a diabetic mum through which I learnt the role of leadership, spontaneous problem solving and staying calm when under pressure. I enjoy baking and sewing, but most of all the art of mehndi, highlighting my manual dexterity and concentration skills. Beautifying women through manicures during a charity event provided a sense of achievement when the ladies left with a smile.

I am a conscientious, individualistic, self critical student with an altruistic attitude. The feeling of helping others is uplifting and to be able to achieve this everyday in the future is an exciting prospect. I am positive that I would be an active addition to university life and I have the capacity to give my all to a challenging dental degree. What better gift to give someone, than the relief of pain and the gift of confidence? I know one day I can do just that, if afforded the opportunity, fulfilling my ambition of becoming an exceptional dental professional.

Universities applied to:

- ➢ Birmingham: Offer
- ➢ Liverpool: Offer
- ➢ Leeds: Offer
- ➢ Kings: Rejection

Good points:

A strong, well-written statement beginning with a nice quote from a famous British philosopher. Already this makes the statement stand out, followed by a deep personal experience to pursue dentistry. Her work experiences cover a wide range of areas in dentistry. An example of relevance further reading to dentistry and would also lead onto obvious questions to be asked at interview.

Bad points:

The statement concentrates a great deal on non academic achievements. There is little to show that the candidate has any real interest or aptitude for the sciences. A dental course is heavily science based as well as practical so it would be sensible to include both aspects.

Overall:

A good statement that is well written and flows easily. It demonstrates a huge amount of experience and shows that they have the skills necessary to be able to cope with the practical side and rigors of a dental course. However a little extra on their academic achievements and interests would have completed the package nicely.

NOTES

Statement 7

There is a certain delight in being naturally curious. Yet this got me in trouble as a child, from asking too many questions to fidgeting to keep my hands busy. In an attempt to nurture my inquisitive character while suppressing my desire to dismantle furniture, I was often encouraged to visit the local museum where my fascination with the osteology of an ancient carnivore led to my discovery of the gargantuan carnassial teeth, fuelling my primitive interests in the morphology and function of teeth. As I grew older, reading texts like 'The Health Gap' fired a passion to engage in the ordeals of social justice and the issue of poor oral health within the NHS, developing my first taste of what would become a fascination with dentistry.

Witnessing the inner workings of NHS practices in areas of high dental need over a week was eye-opening. The sheer variety of cases piqued my interest; allowing me to realise that dentistry is both a stimulating and demanding vocation that is in turn, highly rewarding. The attention to detail taken while placing a filling highlighted that dentistry requires a substantial level of manual dexterity as well as precision and flair. Heading the Dental Society hones these skills, practising needlework to develop dexterity and discussing pertinent dental cases to increase exposure to the field. The fitting of a CEREC crown during a one week placement at a cosmetic practice opened my eyes to technological advancements in the field, prompting further research into possible future innovations.

Completing a Discover Dentistry course placed what I had learnt in dental practice into the wider context of public dental health issues. A culmination of these valuable experiences highlighted that both the beauty and triviality of dentistry lie in the nature of a simple smile; an often overlooked hallmark of social interaction. Shadowing dentists over two weeks during the Goodwill Ambassador Programme offered a striking contrast to previous placements, broadening the parameters of dentistry as a profession that is not only restorative or aesthetic but potentially life saving. While observing the care of a trigeminal neuralgia patient, I was truly able to value the importance of patient autonomy and trust; further affirming that dentistry truly touches lives on a massive scale. Working as part of a multidisciplinary team in a dental hospital highlighted the level of effective communication required in the profession, urging me to draw parallels while managing the Debate Society at college. The methodical nature of the surgical team under the oral surgeon's guidance while treating a motor vehicle trauma patient was provoking. Besides the need for efficient communication, it was clear to me that leadership and management skills were vital; skills that I too, have developed through the Silver DofE Award and leading my team through Young Enterprise. Being a scholarship recipient constantly pushes the horizons of my academic abilities, nurturing my thirst for knowledge and fuelling rigorous self motivation.

Beyond academia, I lead a local charity tutoring disadvantaged children which has given me a deep grounding in community work; stressing the significance of continued community care as well as the values of patience and trust when working with children. Being awarded Best Speaker at the Welsh Debating Championships and being invited to speak at the MDA Awards has fostered an articulate character with the ability to think quickly; making critical decisions under tremendous pressure.

Looking back, it was my curious nature towards the world around me that drove me to explore a career in such a complex and multifaceted field. Grasping every experience extended to me with the same open minded perspective has encouraged me to constantly broaden the frontiers of my perception of dentistry; a vocation that is highly challenging yet calls to me as one that will fulfil my desire to truly make a difference in society.

Universities applied to:

➢ Newcastle: Offer
➢ Birmingham: Offer
➢ Queens University Belfast: Offer
➢ King's College London: Offer with Scholarship

Good points:

The student's opening narrative is not unpleasant to read and is instantly engaging – when done well a personal touch like this can be very effective! Throughout the personal statement the student demonstrates a clear passion for the subject with numerous examples. Moreover, this is evidenced with numerous accounts of clinical exposure and relevant work experience. This is clearly a very academic student with numerous references to significant extracurricular dentistry activities that demonstrate a commitment to the specialty.

Bad points:

Whilst this personal statement is filled with work experience and insight into the dentistry profession, there is little mentioned of the student's personal life. Even though in the penultimate paragraph they start 'beyond academia…', there is no mention of hobbies or relaxation. The examples are all very much academic in nature. When writing a personal statement for a course like dentistry it is essential to demonstrate an interest in a wide variety of unrelated hobbies given the high demands of such a course.

Overall:

An above average statement demonstrating significant insight and commitment within the field of dentistry, written in an engaging synoptic style. Let down, perhaps, only by the lack of an obvious logical structure and neglect of any hobbies or sports.

NOTES

THE UKCAT

The Basics

What is the UKCAT?

The United Kingdom Clinical Aptitude Test (UKCAT) is a two hour computer based exam that is taken by applicants to dental and medical schools. The questions are randomly selected from a huge question bank. Since every UKCAT test is unique, candidates can sit the UKCAT at different times. There is a three month testing period and you can sit the test anytime within it.

You register to sit the test online and book a time slot. On the day, bring along a printout of your test booking confirmation and arrive in good time. Your identity will be checked against a photographic ID that you'll need to bring. You then leave your personal belongings in a locker and enter the test room. Make sure you go to the toilet and have a drink before going in, to save wasting time during the test.

Test Structure:

SECTION	WHAT DOES IT TEST?	QUESTIONS	TIMING
ONE	**Verbal Reasoning**: Essentially a reading comprehension test, with questions based on passages. This is a test of accuracy and speed of reading.	44	22 minutes
TWO	**Decision Making:** This section is brand new for the UKCAT this year, replacing the old decision analysis code section.	29	32 minutes
THREE	**Quantitative Reasoning**: The maths is usually intellectually straightforward but can involve complex or multi stage calculations against the clock. Assesses your understanding of numbers.	36	25 minutes
FOUR	**Abstract Reasoning**: Matching shapes to which set they belong in. This tests your pattern recognition skills and rewards those who have clear and logical organisation of thought.	55	14 minutes
FIVE	**Situational Judgement**: This tests your ability to make decisions in a clinical environment. To perform well, you must understand the role and responsibilities of a dental student.	67	27 minutes

Who has to sit the UKCAT?

You have to sit UKCAT if you are applying for any of the universities that ask for it in the current application cycle. The following is a list of the universities and courses requiring the UKCAT for 2019 entry in the UK taken from the official UKCAT website. At the time of going the press, the details have been updated but not finally confirmed. It is a fairly extensive list, so most applicants will require the UKCAT for one or more of their choices. It would be wise if you are unsure to contact the relevant university admissions to double check.

University	UCAS Course Code
University of Aberdeen	A100, A201
Anglia Ruskin University	A100
Aston University	A100
University of Birmingham	A100, A200
University of Bristol	A100, A108, A206, A208
Cardiff University	A100, A104, A200, A204
University of Dundee	A100, A104, A200, A204
University of East Anglia	A100, A104
University of Edinburgh	A100
University of Exeter	A100*
University of Glasgow	A100, A200
Hull York Medical School	A100
Keele University	A100*, A104*
King's College London	A100, A101, A102, A202, A205, A206
University of Leicester	A100, A199
University of Liverpool	A100*, A200, A201
University of Manchester	A104, A106, A204, A206
University of Newcastle	A100, A101, A206
University of Nottingham	A100, A108
Plymouth University	A100*, A206*
Queen Mary University of London	A100, A101, A110, A120, A130, A200, B960
Queen's University Belfast	A100, A200*
University of Sheffield	A100, A200
University of Southampton	A100, A101, A102
University of St Andrews	A100, A990
St George's, University of London	A100
University of Sunderland	A100 (subject to GMC approval)
University of Warwick	A101

* Alternative requirements may apply to certain groups of students - please see the university website for details

For dentistry there are currently 16 universities that offer the course. Ranking tables may give you an indication about which courses are the best but that does not necessarily mean that the top ranking one will be the correct one for you. Each university has their own merits and the way they run their course will be different. Luckily all the dental schools in the UK are at a high standard so, no matter which one you choose, you will find that the knowledge base is very similar. Therefore it is important to pick the one right for you.

When do I sit the UKCAT?

When you register to sit UKCAT online, you choose your date and time slot and also the test centre. The UKCAT can be sat from 1st July through to early October. Registration for the test opens in early May – we recommend you book your test early so you have the best choice of possible dates.

Many students find it helpful to sit UKCAT in mid-late August – this gives you time in the summer to prepare, but gets the test complete before you go back to school, so you have one less thing to worry about at that busy time. Remember that **you may want to modify your university choices** based on your UKCAT score to maximise your chances of success.

Registration opens	Testing begins	Registration closes	Testing finishes
1 May 2018	2 July 2018	1 Oct 2018	2 Oct 2018

How much does it cost?

In the EU, tests sat in July and August cost £65. Tests sat in September and October cost £85. Tests outside the EU cost £115 throughout the testing period.

Some candidates who might struggle to fund the UKCAT fee are eligible for a bursary. If eligible to apply for one, you need to apply with supporting evidence by the deadline of 21 September 2017.

Bursary Eligibility Criteria

➢ Receipt of 16–19 bursary or EMA.
➢ Receipt of discretionary learner support.
➢ Receipt of full maintenance grant or special support grant.
➢ Receipt of income support, job seeker's allowance or employment support allowance.
➢ Receipt of universal credit.
➢ Receipt of child tax credit.
➢ Receipt of free school meals.
➢ Receipt of EU means tested support.
➢ Living with a family member in support of income support, job seeker's allowance or child tax credit.

Can I re-sit the UKCAT?

You cannot re-sit UKCAT in the same application cycle – whatever score you get is with you for the year. That's why it's so important to make sure you're well prepared and ready to perform at your very best on test day.

If I reapply, do I have to resit the UKCAT?

If you choose to re-apply the following year, you need to sit UKCAT again. You take your new score with you for the new applications cycle. UKCAT scores are only valid for one year from the test date.

When do I get my results?

Because the test is computerised, results are generated immediately and you will be given your score on the day of the test. You will be given a printed sheet with your details and your score to take away. Knowing your score is useful as it can help you choose your universities tactically to maximise your chances of success. Note that you don't put your UKCAT score anywhere on your UCAS form, nor do you contact any universities to inform them. Universities that request UKCAT are sent your scores directly by UKCAT, so you don't need to do anything besides apply through UCAS.

Where do I sit the UKCAT?

UKCAT is a computerised exam and is sat at computer test centres, similar to the driving theory test. When you book the test, you choose the most convenient test centre to sit it at.

How is the UKCAT Scored?

When you finish the test, the computer works out your raw score by adding up your correct responses. There are no mark deductions for incomplete or incorrect answers, so it's a good idea to answer every question even if it's a guess. For the four cognitive sections, this is then scaled onto a scale from 300 – 900. The totals from each of these sections are added together to give your overall score out of 3,600. The new decision making section is now fully introduced into the test, and is scored as normal.

For section 5 (the situational judgement test, SJT) the scoring is slightly different. Here the appropriateness of your responses is used to generate a banding, from band 1 (being the best) to band 4 (being the worst). This is presented separately to the numerical score, such that every candidate's score contains a numerical score out of 3,600 and an SJT banding.

How does my score compare?

This is always a tough question to answer, but it makes sense to refer to the average scores. The scaling is such that around 600 represents the average score in any section, with the majority of candidates scoring between 500 and 700. Thus a score higher than 700 is very good and a score less than 500 is very weak.

For reference, in the 2017 entry cycle the mean scores at the end of testing were 573, 690 and 630 for sections 1, 3 and 4 respectively (section 2 was not marked, but this year it will be), giving an overall score of 1,893 or a mean score of 631 per section.

How is the UKCAT used?

Different universities use your UKCAT score in different ways. Firstly, universities that do not explicitly subscribe to UKCAT cannot see your UKCAT score and are unaware whether or not you sat UKCAT. Universities that use UKCAT can use it in a variety of ways – some universities use it as a major component of the assessment such as selecting candidates for interview based upon the score. Others use it as a smaller component, for example to settle tie-breaks between similar candidates. Each university publishes guidance on how they use the UKCAT, so you should check this out for the universities you are considering.

It's important to know how UKCAT is used in order to maximise your chances. If you score highly in UKCAT, you might decide to choose universities that select for interview based on a high UKCAT score cut-off. That way, you help to stack the odds in your favour – you might, for instance, convert a one in eight chance to a one in three chance. If your score isn't so good, consider choosing universities that don't use UKCAT in that way, otherwise you risk falling at the first hurdle and never getting the chance to show them how great you actually are.

By this logic, **it makes sense for all dental applicants to sit UKCAT** – if you score well it opens doors, and if you don't you don't even have to apply to UKCAT universities. It makes sense not to place all your eggs in one basket.

Can I qualify for extra time?

Yes – some people qualify for extra time in the UKCAT, sitting what is known as UKCAT SEN. If you usually have extra time in public exams at school, you are likely to be eligible to sit the UKCAT SEN. The overall time extension is 30 minutes, bringing the total test time up from 120 to 150 minutes; this is allocated proportionately across the different sections. If you have any dental condition or disability that may affect the test, requiring any special provision, or requiring you to take any dental equipment or medication into the test you should contact customer services to discuss how to best proceed.

Repeat tough questions

When checking through answers, pay particular attention to questions you have got wrong. Look closely through the worked answers in this book until you're confident you understand the reasoning- then repeat the question later without help to check you can now do it. If you use other resources where only the answer is given, have another look at the question and consider showing it to a friend or teacher for their opinion.

Statistics show that without consolidating and reviewing your mistakes, you're likely to make the same mistakes again. Don't be a statistic. Look back over your mistakes and address the cause to make sure you don't make similar mistakes when it comes to the test. You should avoid guessing in early practice. Highlight any questions you struggled with so you can go back and improve.

Positive Marking

When it comes to the test, the marking scheme is only positive – you won't lose points for wrong answers. You gain a mark for each correct answer and do not gain one for each wrong or unanswered one. Therefore if you aren't able to answer a question fully, you should guess. Since each question provides you with 3 to 5 possible answers, you have a 33% to 20% chance of guessing correctly – something which is likely to translate to a number of extra points across the test as a whole.

If you do need to guess, try to make it an educated one. By giving the question a moment's thought or making a basic estimation, you may be able to eliminate a couple of options, greatly increasing your chances of a successful guess. This is discussed more fully in the subsections.

Booking your Test

Unless there are strong reasons otherwise, you should try to **book your test during August**. This is because in the summer, you should have plenty of time to work on the UKCAT and not be tied up with schoolwork or your personal statement deadline. If you book it any earlier, you'll have less time for your all-important preparation; if you book any later, you might get distracted with schoolwork, your personal statement deadline and the rest of your UCAS application. In addition, you pay £20 more to sit the test later in the testing cycle.

More Questions

Practice is the key to UKCAT Success. *The Ultimate UKCAT Guide* contains 1250 Practice questions and covers every section in great detail. You can get a free copy – flick to the back for more details. To practice test timing, there are two full mock papers available at **www.uniadmissions.co.uk/ukcat-mock-papers**.

Verbal Reasoning

The Basics

Section 1 of the UKCAT is the verbal reasoning subtest. It tests your ability to quickly read a passage, find information that is relevant and then analyse statements related to the passage. There are 44 questions to answer and in 21 minutes, so you have just under 30 seconds per question. As with all UKCAT sections, you have one minute to read the instructions. The idea is that this tests both your language ability and your ability to make decisions, traits which are important in a good dentist.

You are presented with a passage, upon which you answer questions. Typically, there are 11 separate passages, each with 4 questions about it. There are two styles of question in section 1, and each requires a slightly different approach. All questions start with a statement relating to something in the passage.

In the first type of question, you are asked if the statement is true or false based on the passage. There is also the option to answer "cannot tell". Choose "true" if the statement either matches the passage or can be directly inferred from it. Choose "false" if the statement either contradicts the passage or exaggerates a claim the passage makes to an extent that it becomes untrue.

Choosing the "cannot tell" option can be harder. Remember that you are answering based *ONLY* on the passage and not on any of your own knowledge – so you choose the "cannot tell" option if there is not enough information to make up your mind one way or the other. Try to choose this option actively. "Cannot tell" isn't something to conclude too quickly, it can often be the hardest answer to select. Choose it when you're actively looking for a certain piece of information to help you answer a question, and you cannot find it.

In the other type of question, you are given a stem and have to select the most appropriate response based on the question. There is only one right answer – if more than one answer seems appropriate, the task is to choose the *best* response. Remember that there is no negative marking in the UKCAT. There will be questions where you aren't certain. If that is the case, then choose an option that seems sensible to you and move on. A clear thought process is key to doing well in section 1 – you will have the opportunity to build that up through the worked examples and practice questions until you're answering like a pro!

This is the first section of the UKCAT, so you're bound to have some nerves. Ensure that you have been to the toilet because once the exam starts you can't pause and go. Take a few deep breaths and calm yourself down. Try to shut out distractions and get yourself into your exam mindset. If you're well prepared, you can remind yourself of that to help keep calm. See it as a job to do and look at the test as an opportunity. If you perform well it will boost your chances of getting into good dental schools. If the worst happens, there are plenty of good dental schools that do not use UKCAT, so all is not lost.

How to Approach This Section

Time pressure is a recurring theme throughout the UKCAT, but it is especially important in Section 1, where you have only 30 seconds per question and a lot of information to take in.

> *Top tip!* Though it might initially sound counter-intuitive, it is often best to read the question *before* reading the passage. When read the passage knowing what you're looking for, you're likely to find the information you need much more quickly.

You should look carefully to see what the question is asking. Sometimes the question will simply need you to find a phrase in the text. In other instances, your critical thinking skills will be needed and you'll have to carefully analyse the information presented to you.

Extreme Words

Words like "extremely", "always" and "never" can give you useful clues for your answer. Statements which make particularly bold claims are less likely to be true, but remember you need a direct contradiction to be able to conclude that they are false.

To answer an "always" question, you're looking for a definition. Always be a bit suspicious of "never" – make sure you're certain before saying true, as most things are possible.

Prioritise

With UKCAT, you can leave and come back to any question. **By flagging for review**, you make this easier. Since time is tight, you don't want to waste time on long passages when you could be scoring easier marks. Score the easy marks first, then come back to the harder ones if time allows. If time runs too short, at least take enough time to guess the answers as there's a good chance you could pick up some marks anyway.

Be a Lawyer

Put on your most critical and analytical hat for section A! Carefully analyse the statements like you're in a court room. Then look for the evidence! **Examine the passage closely, looking for evidence** that either supports or contradicts the statement. Remember **you're making decisions based on ONLY the passage**, not using any prior knowledge. Does the passage agree or disagree? If there isn't enough evidence to decide, don't be afraid to say "cannot tell".

Read the Question First

Follow our top tip and read the question before the passage. There is simply not enough time to read all the passages thoroughly and still have time to complete everything in 22 minutes. By reading the statement or question first, you can understand what it is that is required of you and can then pick out the appropriate area in the passage. Do not fall into the trap of trying to read all of the passage, you will not score highly enough if you do this.

When skim reading through the passage, it is inevitable that you will lose accuracy. However you can reduce this effect by doing plenty of practice so your ability to glean what you need improves. A good tip is to practice reading short sections of complicated texts, such as quality newspapers or novels, at high pace. Then test yourself to see how much you can recall from the passage.

Find the Keywords

The keyword is the most important word to help you relate the question to the passage; sometimes there might be two keywords in a question. When you read the passage, focus in on the keywords straight away. This gives you something to look for in the passage to identify the right place to work from.

It is usually easy to find the keyword/s, and you'll become even better with practice. When you find it, go back a line and read from the line before through the keyword to the end of the line after. Usually, this contains enough relevant information to give you the answer.

If this is not successful, you need to consider your next steps. Time is very tight in the UKCAT and especially so in section 1. There are other passages that need your attention, and there may be much easier marks waiting for you. If reading around the keyword has not given you the right answer it may well be time to move on. It might be that there is a more subtle reference somewhere else, that you need to read the whole passage to reach the answer or indeed that the answer cannot be deduced from the passage. Either way, if it's difficult to find your time could be better spent gaining marks elsewhere. Make a sensible guess and move on.

Use only the Passage

Your answer **must** only be based on the information available in the passage. Do not try and guess the answer based on your general knowledge as this can be a trap. For example, if the question asks who the first person was to walk on the moon, then states "the three crew members of the first lunar mission were Edwin Aldrin, Neil Armstrong and Michael Collins". The correct answer is "cannot tell" – even though you know it was Neil Armstrong and see his name, the passage itself does not tell you who left the landing craft first. Likewise if there is a quotation or an extract from a book which is factually inaccurate, you should answer based on the information available to you rather than what you know to be true.

If you have not been able to select the correct answer, eliminate as many of the statements as possible and guess – you have a 25 – 33% chance of guessing correctly in this section even without eliminating any answers, and if you've read around some keywords in the text you may well have at least some idea as to what the answer is. These odds can add a few easy marks onto your score.

Flagging for Review

There is an additional option to flag a question for review. **Flagging for review has absolutely no effect on the overall score.** All it does is mark the question in an easy way for it to be revisited if you have time later in the section. Once the section is complete, you cannot return to any questions, flagged or unflagged.

Coming back to questions can be inefficient – you have to read the instructions and data each time you work on the question to know what to do, so by coming back again you double the amount of time spent on doing this, leaving less time for actually answering questions. We feel the best strategy is to work steadily through the questions at a consistent and even pace.

That said, flagging for review has one great utility in Section 1. If you come across a particularly long or technical passage, you may want to flag for review immediately and skip on to the next passage. By coming back to the passage at the end, you allow yourself the remaining time on the hardest question. This has an advantage in each of two scenarios. If you're really tight for time, at least you maximised the time you did have answering the easier questions, thereby maximising your marks. If it turns out you have extra time to spare, you can spend it on the hardest question, allowing you a better chance to get marks you otherwise would have struggled to obtain. Thus flagging for review can be useful in Section 1, but its usefulness is probably greatest when you flag questions very soon after seeing them rather than when you have already spent time trying to find the answer.

Remember to find the right balance: if you flag too many questions you will be overloaded and won't have time to focus on them all; if you flag too few, you risk under-utilising this valuable resource. You should flag only a few questions per section to allow you to properly focus on them if you have spare time.

Worked Example

In 287 BC, in the city of Athens, there lived a man named Archimedes who was a royal servant to the King. One day, the King received a crown as a birthday gift and wanted to know whether it was made of pure gold. He ordered Archimedes to find out whether the crown was indeed pure gold or an alloy. For many days, Archimedes pondered over the solution to this problem. He knew the density of gold, but could not calculate the volume of the crown.

One day, as he was bathing, he realised as he got into the bath that the volume of water displaced must be exactly equal to the volume of his own body. Upon this realisation he ran across the streets naked, yelling eureka! He weighed the crown and found its volume by immersing it in water and then calculated its density. He discovered that the density did not match that of pure gold. The crown was impure, and the blacksmith responsible for its manufacture suffered the consequences.

1. Archimedes knew the volume of the crown but could not calculate its weight
    a. True
    b. False
    c. Cannot tell

Look for the keyword!

2. Archimedes gave the crown as a birthday gift to the King
    a. True
    b. False
    c. Cannot tell

THEN search the passage

3. The crown had silver impurities
    a. True
    b. False
    c. Cannot tell

4. Archimedes found the weight of the crown using a balance scale
    a. True
    b. False
    c. Cannot tell

Answers

1. **False** – The keywords are volume and weight. Check these against the text and you will find that Archimedes could calculate the weight, but not the volume.
2. **False** – Whilst it does not explicitly state the giver of the gift, the description of Archimedes as a servant and his role in investigating the crown is wholly incompatible with him being the giver of the gift.
3. **Cannot tell** – The word silver does not appear anywhere in the passage so this statement cannot be true. But this statement is not false either because nowhere does it say that silver was not the impurity.
4. **Cannot tell** – Through your own logic, you probably guessed that this is how Archimedes weighed the crown, but remember to only use the information within the passage and use of a balance scale is not mentioned.

Verbal Reasoning Questions

For questions 1 – 20 decide if each of the statements is true, false or can't tell:

SET 1

The Kyoto Protocol is an international agreement written by the United Nations in order to reduce the effects of climate change. This agreement sets targets for countries in order for them to reduce their greenhouse gas emissions. These gases are believed to be responsible for causing global warming as a result of recent industrialisation. The Protocol was written in 1997 and each country that signed the protocol agreed to reduce their emissions to their own specific target. This agreement could only become legally binding when two conditions had been fulfilled: When 55 countries agreed to be legally bound by the agreement and when 55% of emissions from industrialised countries had been accounted for.

The first condition was met in 2002 however countries such as Australia and the United States refused to be bound by the agreement so the minimum of 55% of emissions from industrialised countries was not met. It was only after Russia joined in 2004 that allowed the protocol to come into force in 2005. Some climate scientists have argued that the target combined reduction of 5.2% emissions from industrialised nations would not be enough to avoid the worst consequences of global warming. In order to have a significant impact, we would need to aim at reducing emissions by 60% and to get larger countries such as the US to support the agreement.

1. The Kyoto Protocol is legally binding in all industrialised countries.
2. The greenhouse gas emissions from Australia and the United States represent 45% of emissions from industrialised countries.
3. Each country chose the amount by which they would reduce their own emissions.
4. The global emission of greenhouse gases has reduced since 2005.
5. The harmful effects of climate change would be avoided if all countries reduced their emissions by 60%.

SET 2

The space race was a competition between the Soviet Union and the United States to show off their technological superiority and economic power. It took place during the Cold War when there was a tense relationship between these nations. As the technology used in space exploration could also have military applications, both nations had many scientists and technicians involved.

In 1957, the USSR launched the first artificial satellite into the Earth's orbit, named Sputnik. The launch of this satellite was one of the first steps towards space exploration. The Americans were worried that the Soviets could use similar technology to launch nuclear warheads. This prompted urgency within the Americans, leading President Eisenhower to found NASA, and so began the space race. The Soviets took another step forward in April 1961 when they sent the first person into space, a cosmonaut named Yuri Gagarin. This prompted President John F. Kennedy to make the unexpected claim that the US would beat the Soviets to land a man on the moon and that they would do so before the end of the decade. This led to the foundation of Project Apollo, a programme designed to do this.

In 1969, Neil Armstrong & Buzz Aldrin set off for the moon on the Apollo 11 space mission and became the first astronauts to walk on the moon. Neil Armstrong famously said "one small step for man, one giant leap for mankind." This lunar landing led the US to win the space race that started with Sputnik's launch in 1957.

6. The Soviet Union were mainly concerned with launching satellites into space for a military advantage over the United States.
7. Project Apollo was founded in order for the United States to defeat the Soviet Union in the Cold War.
8. Yuri Gagarin did not become the first man on the moon because the Soviet technology could not handle the conditions on the moon.
9. The United States began their attempts at space exploration when the Soviets launched Sputnik.
10. The United States were losing the space race when John F. Kennedy said they would land a man on the moon.

SET 3

A marathon is a long distance running event that is 26.2 miles long. This race was named after the famous Battle of Marathon. The first Persian invasion of Greece took place in 490 BC. The Greek soldiers did not expect to defeat the Persian army, which had greater numbers and superior cavalry. The Greek commander utilised a tactical flank to defeat the Persians forcing them to retreat back to Asia. According to legend, the fastest Greek runner, Pheidippides, was ordered to run from Marathon to Athens to announce the Greek victory over the Persians, but then collapsed and died of exhaustion. This legendary 25 mile journey from Marathon to Athens is the basis for modern marathons.

The initial organisers of the Olympic Games in 1896 wanted an event that would celebrate the glory of Ancient Greece. They therefore chose to use the same course that Pheidippides ran. In subsequent Olympic Games, the exact length of the route depended on the location but was roughly similar to the 25 mile distance. The current standardised distance of 26.2 miles has been chosen by the IAAF and used since 1921, and has been taken from the distance used at the 1908 Olympics in London. Nowadays, more than 500 marathons are organised each year.

11. Pheidippides was chosen as he was the only Greek runner determined enough to make the journey to Athens
12. Marathon distances have been standardised since the 1908 Olympics
13. The Persian commander believed he would defeat the Greeks in the Battle of Marathon.
14. The original route from Marathon to Athens is used for IAAF marathons today.
15. The Persian soldiers were trained better than the Greek soldiers.

SET 4

Many species of bird migrate northwards in the spring to take advantage of the abundance of nesting locations and insects to eat. As the availability of food resources decreases during the winter to the point where the birds cannot survive, the birds migrate south again. Some species are capable of flying all the way around the earth.
The act of migration itself can be risky for birds due to the amount of energy required to sustain flight over these long distances. Many juvenile birds can die from exhaustion during their first migration. Due to this inherent risk of migration, many species of birds have acquired different adaptations to increase the efficiency of flight. Flying with other birds in certain formations can allow their flight patterns to be more energy efficient.

The Northern bald ibis migrates from Austria to Italy. The behaviour of these birds is such that they migrate together within a flock and each individual bird continuously changes its position within the flock. Each individual bird benefits by spending some time flying in the updraft produced by the leading birds and a proportional amount of time leading the formation. Although it would theoretically be possible for an individual bird to take advantage of this energy-efficient flight without leading the formation itself, no Northern bald ibis has been shown to do this.

16. As migration is risky and dangerous, it would be better for birds not to migrate.
17. All migrating birds do so in flocks to increase their efficiency
18. All the birds within a flock of Northern bald ibis benefit from flocking behaviour
19. A bird within a Northern bald ibis flock that does not lead will be forbidden from flying with the rest of the flock
20. The migration timing depends on different seasons

SET 5

Geology deals with the rocks of the earth's crust. It learns from their composition and structure how the rocks were made and how they have been modified. It ascertains how they have been brought to their present places and wrought to their various topographic forms, such as hills and valleys, plains and mountains. It studies the vestiges, which the rocks preserve, of ancient organisms that once inhabited our planet. Geology is the history of the earth and its inhabitants, as read in the rocks of the earth's crust.

To obtain a general idea of the nature and method of our science before beginning its study in detail, we may visit some valley, on whose sides are rocky ledges. Here the rocks lie in horizontal layers. Although only their edges are exposed, we may infer that these layers run into the upland on either side and underlie the entire district; they are part of the foundation of solid rock found beneath the loose materials of the surface everywhere.

Take the sandstones ledge of a valley. Looking closely at the rock we see that it is composed of myriads of grains of sand cemented together. These grains have been worn and rounded. They are sorted also, those of each layer being about of a size. By some means they have been brought hither from some more ancient source. Surely these grains have had a history before they here found a resting place—a history which we are to learn to read.

The successive layers of the rock suggest that they were built one after another from the bottom upward. We may be as sure that each layer was formed before those above it as that the bottom courses of stone in a wall were laid before the courses which rest upon them.

21. Based on the passage, each of these statements can be verified, EXCEPT?
A.   We can learn about earth's inhabitants through its crust.
B.   Individual layers of sandstone form one after another.
C.   Rocks are made of sand.
D.   Geology does not always demand explicit evidence.

22. Wall-building is used in this passage to help us understand:
A.   Mountains
B.   Valleys
C.   Hills
D.   Plains

23. The sand mentioned in the passage comes from:
A.   An ancient beach
B.   The sea
C.   The earth's crust
D.   It is undisclosed

24. A foundation of rock is **NOT** found underneath:
A.   Upland
B.   Lowland
C.   Nowhere
D.   Water

25. 'Grains of sand' are described as sorted by:
A.   Shape
B.   Texture
C.   Age
D.   Measurements

SET 6

The genus of plants called Narcissus, many of the species of which are highly esteemed by the floriculturist and lover of cultivated plants, belongs to the Amaryllis family (Amaryllidaceæ.) This family includes about seventy genera and over eight hundred species that are mostly native in tropical or semi-tropical countries, though a few are found in temperate climates.

Many of the species are sought for ornamental purposes and, on account of their beauty and remarkable odour, they are more prized by many than are the species of the Lily family. In this group is classed the American Aloe (Agave Americana) valued not only for cultivation, but also by the Mexicans on account of the sweet fluid which is yielded by its central bud. This liquid, after fermentation, forms an intoxicating liquor known as pulque. By distillation, this yields a liquid, very similar to rum, called by the Mexicans mescal. The leaves furnish a strong fibre, known as vegetable silk, from which, since remote times, paper has been manufactured.
The popular opinion is that this plant flowers but once in a century; hence the name 'Century Plant' is often applied to it, though under proper culture it will blossom more frequently.

26. Which of the following are **NOT** mentioned as potential uses for a narcissus plant:

A.   Perfume production
B.   Alcohol production
C.   Visual decoration
D.   Stationary production

27. Why is the plant known as 'the century plant'?

A.   It is sown only once every hundred years.
B.   It can only able to be fertilised once a century.
C.   It is perceived as blooming centennially.
D.   It can only able to flower once within a hundred years.

28. Which of the following statements is most supported by the above passage:

A.   Lilies are generally valued less than members of the Narcissus genus.
B.   Lilies are famously not as attractive as members of the Narcissus genus.
C.   A number are people prefer members of the Narcissus genus over Lilies.
D.   Members of the Narcissus genus are a welcome addition to any household.

29. Which of the following statements is NOT true:

A.   American Aloe can be used to make rum.
B.   The Amaryllis family contains more than six hundred species of Narcissus.
C.   Members of the Narcissus genus can be found in all climates.
D.   The members of the Narcissus genus have a distinctive smell.

30. Which of the following statements can be verified by the passage:

A.   The 'Narcissus' genus is named after the mythical character, famed for his beauty.
B.   Agave syrup can be collected by American Aloe.
C.   A genus belongs to a family.
D.   Members of the Narcissus genus are used for their soothing properties.

SET 7

The following passage is found in a book on nature published in 1899:

Five women out of every ten who walk the streets of Chicago and other Illinois cities, says a prominent journal, by wearing dead birds upon their hats proclaim themselves as lawbreakers. For the first time in the history of Illinois laws it has been made an offense punishable by fine and imprisonment, or both, to have in possession any dead, harmless bird except game birds, which may be possessed in their proper season. The wearing of a tern, or a gull, a woodpecker, or a jay is an offense against the law's majesty, and any policeman with a mind rigidly bent upon enforcing the law could round up, without a written warrant, a wagon load of the offenders any hour in the day, and carry them off to the lockup. What moral suasion cannot do, a crusade of this sort undoubtedly would.

Thanks to the personal influence of the Princess of Wales, the osprey plume, so long a feature of the uniforms of a number of the cavalry regiments of the British army, has been abolished. After Dec. 31, 1899, the osprey plume, by order of Field Marshal Lord Wolseley, is to be replaced by one of ostrich feathers. It was the wearing of these plumes by the officers of all the hussar and rifle regiments, as well as of the Royal Horse Artillery, which so sadly interfered with the crusade inaugurated by the Princess against the use of osprey plumes. The fact that these plumes, to be of any marketable value, have to be torn from the living bird during the nesting season induced the Queen, the Princess of Wales, and other ladies of the royal family to set their faces against the use of both the osprey plume and the aigrette as articles of fashionable wear.

31. In 1899:
A. Women across the USA could be prosecuted for owning ornamental dead birds.
B. There was a significant rise of female arrests in America.
C. Possession of a dead gull could lead to trouble.
D. Americans responded to law by citing the use of jays as ornamentation unfashionable.

32. Ostrich feathers were seen as preferable to osprey plums because:
A. Ostriches are less intelligent birds.
B. Ostriches are killed for their meat, so one might as well use their feathers.
C. Queen Elizabeth has an especial love of ospreys.
D. Harvesting osprey feathers was seen as an inhumane process.

33. Games birds could be possessed by citizens of Illinois all year round.
A. True
B. False
C. Can't tell

34. Banning Osprey feathers in the UK's army was difficult because:
A. Many uniforms required them.
B. The Princess did not have the authority to implement the ban.
C. Her ultimate support was predominately female, and thus their concerns seemed to have no relevance from the male domain of the army.
D. It would be hard to differentiate between other regiments within the army, who were already wearing ostrich feathers.

35. Which of the following could NOT be legally owned in Illinois, according to the passage:
A. A live bird intended for personal ornamentation.
B. A dead bird of prey that had violently attacked you.
C. Feathered garments.
D. None of the above.

Decision Making

The Basics

The Decision Making section is **brand new** to the UKCAT in 2017. It replaces an old section called Decision Analysis which involved deciphering the meaning of various coded phrases and is no longer tested.

The section lasts 31 minutes (with one additional minute to read the instructions), and there are 29 questions to be answered – so you need to work quickly and efficiently at a rate of about one question per minute. You will be presented with questions that may refer to text, charts, tables, graphs or diagrams. All of the questions are standalone and do not share data, so make sure to focus on each question independently.

This section was a component of the test last year, but was not scored. This year the section will be scored just like every other subsection, and the score from this section will contribute to your overall score and result.

The Questions

The idea behind this section is to assess how you use information and data to make decisions – a skill that is essential to working effectively as a dentist. The questions come in a variety of styles, but all are focussed on testing your decision making ability.

The questions making up the decision making section can be broken down into six main styles. All of the questions in this section will belong to one of these styles, so by familiarising yourself with the theory you will make it much easier to answer the questions. The six styles are:

1) **Logical Puzzles**
2) **Syllogisms**
3) **Interpreting Information**

4) **Recognising Assumptions**
5) **Venn Diagrams**
6) **Probabilistic Reasoning**

The questions may provide multiple pieces of information which build together to give the overall picture. Remember to consider each of these in turn and build up your understanding of the situation in pieces before bringing it all together.

The most important thing is to understand the premise of the question. There is often a clear chain of logic from the start of the question to the final answer, so the sooner you work this out, the sooner you can begin answering on the right track.

Strategy

As there are different styles of questions in this section, the best way to think about preparation is to subdivide the questions into their different styles and develop a clear approach to each

As a general principle, drawing diagrams can be helpful. If for example, the question has a number of people in it, write their names down on your whiteboard (or just the first initial to save space and time). If the question talks about compass directions, draw a simple map. If it talks about categories, a Venn diagram can help. If it involves a timetable or timed events, draw out a timeline to simplify your thought process. By using visual tools to complement the words, you make it easier to understand and solve the problem.

Logical Puzzles

The logical puzzle questions require you to make an inference based on the available information to get to the answer. Commonly, this will manifest in being presented with some background information (a general statement) and then some extra information (a specific statement), and both must be combined to make the conclusion and find the answer. Deductive reasoning can be used in either a positive way to prove something, or in a negative way to disprove something. By familiarising yourself with the structures, you will improve your ability to notice and use the relevant information.

An example of **positive inductive reasoning**:
- i. All birds have wings
- ii. Ostriches are birds
- iii. Therefore the ostrich has wings

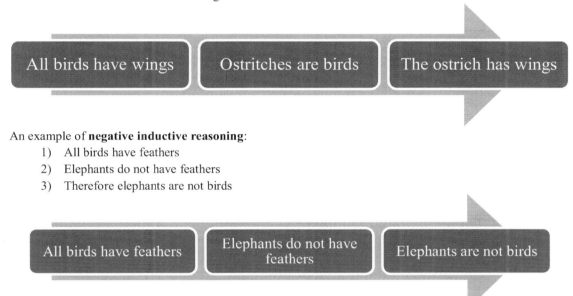

An example of **negative inductive reasoning**:
1) All birds have feathers
2) Elephants do not have feathers
3) Therefore elephants are not birds

Syllogisms

Syllogisms are an additional application of deductive reasoning. In these questions, you will be presented with information that can be used to make certain conclusions, but which will also give and incomplete description of the situation. Then, the task is to determine which answer/answers is/are supported and which aren't. To do well in these questions, you have to be very clear about the limitations of the information available – if the statement cannot be deduced from the information in the questions, then it is not true.

Interpreting information

For the interpreting information questions, you will be presented with a more complex and less directly relevant set of information than in the logical deduction questions. This information may be in the form of a passage of text describing something, or alternatively it could be in the form of a table, chart or graph. You will then have to use the information source to extract the relevant information to answer the question. Don't be afraid to use rounding and estimations – if the differences are substantial, you may not need to calculate figures exactly.

To answer these questions effectively, look at the question before digesting the data. Once you understand the premise of the question, you can approach the data in a much more focussed way to gather the information you need and ignore the distracters designed to make the question more difficult. When analysing graphs and charts, always follow a systematic approach to ensure you grasp the key message as easily as possible, such as the method illustrated below.

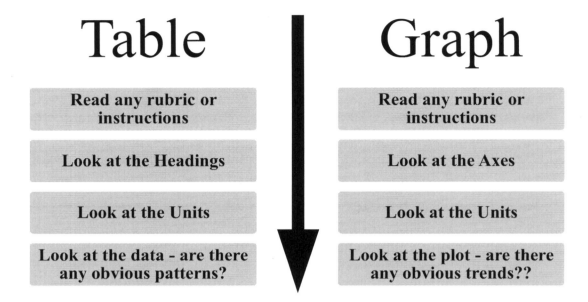

Recognising assumptions

Assumptions are an important component of decision making, and proper use of them is essential to being a good decision maker. Relying too heavily on assumptions leaves the decision at risk of being wrong, whereas reluctance to make assumptions can make the decision process extremely slow and laborious. The questions of this style aim to probe your understanding of assumption making as a component of the decision process.

Many of these questions will ask you to select the strongest argument for or against a statement. To help you select the best option, use the acronym **FREES**. Factual – the argument should be based on fact rather than opinion. Relevant – the argument should directly address the statement in the question. Entire – the argument should address the whole question, not only one aspect of it. Emotionless – the argument should avoid emotional pleas and derive strength from relevant evidence. Sensible – the argument should be a generally sensible and reasonable approach to take.

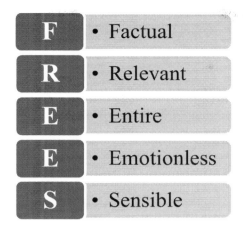

Venn diagrams

The category of "Venn diagram" questions encompasses any that require the use of Venn diagrams within the question. Venn diagrams might be used in the question itself to present the data, they may be required as a part of the working to deduce the correct answer, or it may be that the answers are presented in the form of Venn diagrams, and you have to choose the most appropriate response.

Venn diagrams can take a number of forms. Some may look like the style of diagram you will have seen for many years, with two or three overlapping segments into which items are sorted. But they don't necessarily look like that. Provided the basic rules are followed – that to sort any item, it is placed into each and every circle that it belongs to – then any shape can be formed. The shape and structure of the diagrams has to be altered to allow all circles that need to overlap to be able to do so. To build your skills, practice drawing Venn diagrams to classify common objects – for example vehicles, kitchen utensils, farm animals or school subjects in different ways. Below are examples of Venn diagram structures that you may encounter in this section of the UKCAT.

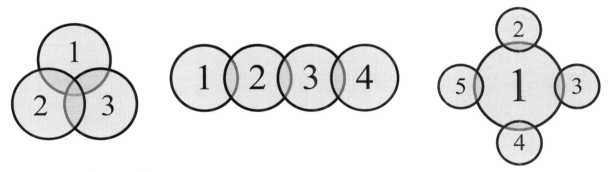

Probabilistic reasoning

These questions assess your ability to use probabilistic reasoning in the decision making process. You may be presented with probabilistic information in a variety of forms – fractions, decimals, percentages or odds – so do remind yourself of these different notations if you haven't seen them in a while. Whenever you see a probability, take the time to note exactly what occurrence the probability is representing, and whether it is the positive probability (of a thing happening) or a negative probability (of it not happening).

A good rule of thumb with probability is this. If you are asked the probability of something **or** something occurring, then the overall probability is higher than only one of them occurring so you add the probabilities. If you are asked the probability of something **and** something occurring, then the overall probability is lower than only one of them occurring, so you multiply the probabilities. For example, if asked about the probability of rain on two given consecutive days, when the probability of rain on any given day is 0.4, then it can be calculated as follows:

$$\boxed{\text{Dry} = 0.6}$$

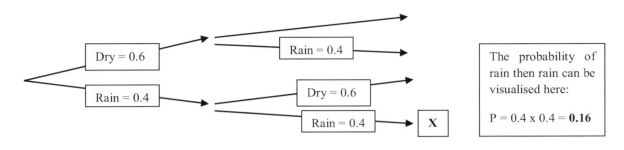

The probability of rain then rain can be visualised here:

P = 0.4 x 0.4 = **0.16**

Top tip! If you are ever confused about probability, draw out a probability tree. The visualisation will enable you to see through the complexity to identify the correct calculation.

Worked examples

Example 1

A Surgeon is planning his surgery list for the next day. He has a total of 5 surgeries planned for the day. Mrs Smith has an infected arm wound that needs to be cleaned and stitched. Due to the infection, the surgeon will operate on her last.

> Start building up the order as information in gathered

Mr Hunt has a broken leg and will be operated on before Mr Pierce.
Mr Perry will be operated on first, his procedure will take 1hr.
Mr Dutch has his procedure 2hrs before Mr Hunt and 60mins after Mr Perry.

Which of the following statements is **TRUE**?

> To be true, we need to see DIRECT supporting evidence

A. Mr Hunt's leg is infected
B. Mr Dutch will be operated on third
C. There are no women on the operating list
D. **Mr Pierce will be operated on after Mr Hunt**

Answer **D**: the only order that satisfies the information is Perry-Dutch-Hunt-Pierce-Smith

Example 2

A group of scientists conduct research into the social structure of Vikings. They find that Vikings tended to be organised in close knit groups of families often living in individual villages. Leadership was often organised along performance as a leader and rarely was a birth right like in other early mediaeval societies in central Europe. Slavery was a common practice in Viking society with slaves coming from all areas raided by individual groups or from centralised slave markets.

> This is a syllogism style – you need to work out which answers are supported by the information in the text and which ones are not. This may involve logical reasoning.

Which of the following statements is **INCORRECT**?

A. Central European societies were controlled by birth right lordship
B. **Viking society was democratic**
C. The slave trade continued to be present in medieval societies
D. Viking society was close knit

Answer **B**: There is no mention of democracy; all other answers are supported by the text

> ***Top tip!*** Read the question before starting to interpret tables, charts or graphs. That way, you know what information you need and what is there to distract you!

Decision Making Questions

Question 1:

Pilbury is south of Westside, which is south of Harrington. Twotown is north of Pilbury and Crewville but not further north than Westside. Crewville is:

A. South of Westside, Pilbury and Harrington but not necessarily Twotown.
B. North of Pilbury, and Westside.
C. South of Westside and Twotown, but north of Pilbury.
D. South of Westside, Harrington and Twotown but not necessarily Pilbury.
E. South of Harrington, Westside, Twotown and Pilbury.

Question 2:

The hospital coordinator is making the rota for the ward for next week; two of Drs Evans, James and Luca must be working on weekdays, none of them on Sundays and all of them on Saturdays. Dr Evans works 4 days a week including Mondays and Fridays. Dr Luca cannot work Monday or Thursday. Only Dr James can work 4 days consecutively, but he cannot do 5.

What days does Dr James work?

A. Saturday, Sunday and Monday.
B. Monday, Tuesday, Wednesday, Thursday and Saturday.
C. Monday, Thursday Friday and Saturday.
D. Tuesday, Wednesday, Friday and Saturday.
E. Monday, Tuesday, Wednesday, Thursday and Friday.

Question 3:

If criminals, thieves and judges are represented below:

Assuming that judges must have clean record, all thieves are criminals and all those who are guilty are convicted of their crimes, which of one of the following best represents their interaction?

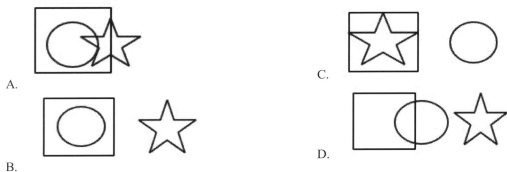

A.

B.

C.

D.

Question 4:

Apples are more expensive than pears, which are more expensive than oranges. Peaches are more expensive than oranges. Apples are less expensive than grapes.

Which two of the following must be true?

A. Grapes are less expensive than oranges.
B. Peaches may be less expensive than pears.
C. Grapes are more expensive than pears.

D. Pears and peaches are the same price.
E. Apples and peaches are the same price.

Question 5:

A class of young students has a pet spider. Deciding to play a practical joke on their teacher, one day during morning break one of the students put the spider in their teachers' desk. When first questioned by the head teacher, Mr Jones, the five students who were in the classroom during morning break all lied about what they saw. Realising that the students were all lying, Mr Jones called all 5 students back individually and, threatened with suspension, all the students told the truth. Unfortunately Mr Jones only wrote down the student's statements not whether they had been told in the truthful or lying questioning.

The students' two statements appear below:

Archie: "It wasn't Edward. "
        "It was Bella."

Charlotte: "It was Edward."
            "It wasn't Archie"

Darcy: "It was Charlotte"
        "It was Bella"

Bella: "It wasn't Charlotte."
        "It wasn't Edward."

Edward: "It was Darcy"
        "It wasn't Archie"

Who put the spider in the teacher's desk?

A.   Edward
B.   Bella
C.   Darcy

D.   Charlotte
E.   More information needed.

Question 6:

On a specific day at a GP surgery 150 people visited the surgery and common complaints were recorded as a percentage of total patients. Each patient could use their appointment to discuss up to 2 complaints. 56% flu-like symptoms, 48% pain, 20% diabetes, 40% asthma or COPD, 30% high blood pressure.
Which statement must be true?

A. A minimum of 8 patients complained of pain and flu-like symptoms.
B. No more than 45 patients complained of high blood pressure and diabetes.
C. There were a maximum of 21 patients who did not complain about flu-like symptoms or high blood pressure.
D. There were actually 291 patients who visited the surgery.
E. None of the above.

Question 7:

During a GP consultation in 2015, Ms Smith tells the GP about her grandchildren. Ms Smith states that Charles is the middle grandchild and was born in 2002. In 2010, Bertie was twice the age of Adam and that in 2015 there are 5 years between Bertie and Adam. Charles and Adam are separated by 3 years.

How old are the 3 grandchildren in 2015?

A. Adam = 16, Bertie = 11, Charles = 13

B. Adam = 5, Bertie = 10, Charles = 8

C. Adam = 10, Bertie = 15, Charles = 13

D. Adam = 10, Bertie = 20, Charles = 13

E. Adam = 11, Bertie = 10, Charles = 8

F. More information needed.

Question 8:

A team of 4 builders take 12 days of 7 hours work to complete a house. The company decides to recruit 3 extra builders. How many 8 hour days will it take the new workforce to build a house?

A. 2 days

B. 6 days

C. 7 days

D. 10 days

E. 12 days

F. More information needed

Question 9:

Four young girls entered a local baking competition. Though a bit burnt, Ellen's carrot cake did not come last. The girl who baked a Madeira sponge had practiced a lot, and so came first, while Jaya came third with her entry. Aleena did better than the girl who made the Tiramisu, and the girl who made the Victoria sponge did better than Veronica.

Which **TWO** of the following were **NOT** results of the competition?

A. Veronica made a tiramisu

B. Ellen came second

C. Aleena made a Victoria sponge

D. The Victoria sponge came in 3rd place

E. The carrot cake came 3rd

Question 10:

John likes to shoot bottles off a shelf. In the first round he places 16 bottles on the shelf and knocks off 8 bottles. 3 of the knocked off bottles are damaged and can no longer be used, whilst 1 bottle is lost. He puts the undamaged bottles back on the shelf before continuing. In the second round he shoots six times and misses 50% of these shots. He damages two bottles with every shot which does not miss. 2 bottles also fall off the shelf at the end. He puts up 2 new bottles before continuing. In the final round, John misses all his shots and in frustration, knocks over gets angry and knocks over 50% of the remaining bottles.

How many bottles were left on the wall after the final round?

A. 2

B. 3

C. 4

D. 5

E. 6

F. More information needed.

Question 11:

A bus takes 24 minutes to travel from White City to Hammersmith with no stops. Each time the bus stops to pick up and/or drop off passengers, it takes approximately 90 seconds. This morning, the bus picked up passengers from 5 stops, and dropped off passengers at 7 stops. What is the minimum journey time from White City to Hammersmith this morning?

A. 28 minutes   C. 34.5 minutes   E. 37.5 minutes
B. 34 minutes   D. 36 minutes   F. 42 minutes

Question 12:

I look at the clock on my bedside table, and I see the following digits:

However, I also see that there is a glass of water between me and the clock, which is in front of 2 adjacent figures. I know that this means these 2 figures will appear reversed. For example, 10 would appear as 01, and 20 would appear as 05 (as 5 on a digital clock is a reversed image of a 2). Some numbers, such as 3, cannot appear reversed because there are no numbers which look like the reverse of 3.

Which of the following could be the actual time?
A. 15:52   D. 12:22
B. 21:25   E. 21:52
C. 12:55

Question 13:

Ryan is cooking breakfast for several guests at his hotel. He is frying most of the items using the same large frying pan, to get as much food prepared in as little time as possible. Ryan is cooking Bacon, Sausages, and eggs in this pan. He calculates how much room is taken up in the pan by each item. He calculates the following:
-   Each rasher of bacon takes up 7% of the available space in the pan
-   Each sausage takes up 3% of the available space in the pan.
-   Each egg takes up 12% of the available space in the pan.

Ryan is cooking 2 rashers of bacon, 4 sausages and 1 egg for each guest. He decides to cook all the food for each guest at the same time, rather than cooking all of each item at once.

How many guests can he cook for at once?

A. 1
B. 2
C. 3
D. 4
E. 5

Question 14:

Northern Line trains arrive into Kings Cross station every 8 minutes, Piccadilly Line trains every 5 minutes and Victoria Line trains every 2 minutes. If trains from all 3 lines arrived into the station exactly 15 minutes ago, how long will it be before they do so again?

A. 24 minutes   C. 40 minutes   E. 65 minutes
B. 25 minutes   D. 60 minutes   F. 80 minutes

Question 15:

In how many different positions can you place an additional tile to make a straight line of 3 tiles?

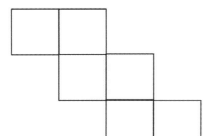

A. 6   D. 9
B. 7   E. 10
C. 8   F. 11

Question 16:

Ellie, her brother Tom, her sister Georgia, her mum and her dad line up in height order from shortest to tallest for a family photograph. Ellie is shorter than her dad but taller than her mum. Georgia is shorter than both her parents. Tom is taller than both his parents. If 1 is shortest and 5 is tallest, what position is Ellie in the line?

A. 1   D. 4
B. 2   E. 5
C. 3

Question 17:

Miss Briggs is trying to arrange the 5 students in her class into a seating plan. Ashley must sit on the front row because she has poor eyesight. Danielle disrupts anyone she sits next to apart from Caitlin, so she must sit next to Caitlin and no-one else. Bella needs to have a teaching assistant sat next to her. The teaching assistant must be sat on the left hand side of the row, near to the teacher. Emily does not get on with Bella, so they need to be sat apart from one another. The teacher has 2 tables which each sit 3 people, which are arranged 1 behind the other. Who is sitting in the front right seat?

A. Ashley   D. Danielle
B. Bella   E. Emily
C. Caitlin

Question 18:

Piyanga writes a coded message for Nishita. Each letter of the original message is coded as a letter a specific number of characters further on in the alphabet (the specific number is the same for all letters). Piyanga's coded message includes the word "PJVN". What could the original word say?

A. CAME   D. GAME
B. DAME   E. LAME
C. FAME

For 80 more DM questions check the *Ultimate UKCAT Guide*– flick to the back to get a free copy.

Quantitative Reasoning

The Basics

The Quantitative Reasoning subtest tests your ability to quickly interpret data and perform relevant calculations upon them. Section 3 contains 36 questions and you have 24 minutes to answer them, giving a total of 40 seconds per question – slightly more generous than the verbal reasoning section.

There are different types of question you can be asked in Section 3, but all involve interpreting a numerical data source and performing calculations. This is all about testing your natural ability with numbers, how easily you understand numbers and how well you can make calculations based upon new data. You won't find advanced mathematics, so you are at absolutely no disadvantage by not taking A-level maths. Common sources include food menus, timetables, sales figures, surveys, conversion tables and more. In this section, the whiteboard you are provided can be useful – use it to scribble down working and intermediate numbers as required.

There is an on screen calculator – a basic calculator for performing arithmetic. You should ideally **practice with a non-scientific calculator** when working through this book, as that will give the closest simulation to what you will get on the day. When you move on to trying the online UKCAT practice papers, the calculator is available on screen as you will see it in the test. In addition to using it to solve questions, practice different calculations to build your speed using it – though this sounds boring, it will save you valuable time on the day. Something many candidates do not realise is that the calculator can be operated using the keyboard controls. Try this out for yourself in practice, and if it works for you it is yet another way to boost your speed when you come to the UKCAT for real.

Preparation

Be comfortable using the on-screen calculator
As discussed, it's important to know exactly how the calculator works so that you use it quickly and most effectively during the test. Ensure you know what functions the calculator has, and does not have. Use the automated practice section of the UKCAT website for this – that way, you practice using the same software you will use on the day.

Practice common question styles
Be especially comfortable with things like bus and rail timetables, sales figures, surveys, converting units and working with percentage changes in both directions. These are commonplace in the UKCAT –, but could prove awkward if you're rusty. Likewise be sharp on your simple arithmetic – it might seem basic, but a good knowledge of times tables will save you a lot of time. Even if you're not answering questions, you can hone your skills by practicing reading charts, graphs and tables quickly.

Familiarise yourself with the format of diagrams
Working through plenty of practice questions will help here, as you'll see similar questions coming up again and again. Commonly you will need to use timetables, data tables and different types of graphs to answer Section 3 questions. Make sure you are comfortable with all of these styles of questions.

Mental Speed
The main challenge in most questions is finding the right data and selecting the appropriate calculation to perform, rather than the actual calculation itself. However, time is tight so you should be confident with addition, subtraction, multiplication, division, as well as working out percentages, fractions and ratios. Although there is an on-screen calculator in the test, you can save time by doing the basic sums in your head – being confident in your mental arithmetic ability will help you use the time most effectively. There are some good websites and apps to practice quick-fire mental arithmetic – using these you can quickly refresh these essential skills.

Answering Questions

Estimation

Estimation can be very helpful, particularly when the answers are significantly different. If, for instance, answers are an order of magnitude or more away from each other, you can ignore the fine print of the numbers and still get the right answer. If it's a particularly complicated calculation, quickly ask yourself roughly what answer you are expecting. With the simple UKCAT calculator it can be easy to slip up – but if you've already made a quick estimation then you may be alerted to your mistake before you put the wrong answer down. Another important use of estimation is to generate educated guesses if you're short on time. In Section 3 there are five answers per question, so your odds of blindly guessing correctly are low. But here a simple estimation can help. A quick glance or simplified sum might help you eliminate a few answers in only seconds, boosting the chance your guess is correct.

Flagging for review

Flagging for review is so quick and easy, it can always be a useful tool. If you're finding a question difficult, or you've decided it is likely to take too long to solve, put a guess (or quick estimation if possible), flag for review and move on. This allows you to revisit the question at the end if there's time whilst using your time more efficiently elsewhere. When doing this you should make an initial guess, as this ensures you have at least a chance of being correct if you don't have enough time to come back again.

Pace yourself

In this section you have an average of 40 seconds per question, and this is a very useful guide to have. Of course some questions will take more or less time, but you should aim to work steadily forwards at roughly that pace. So after 6 minutes you should be about 9 questions in, and after 12 minutes should have completed about 18 questions and so on. By keeping a regular rhythm to your work, you ensure you don't leave lots of potentially easy questions at the end untouched. It's far better to skip a few tricky questions with a guess to make sure you make a decent effort at all questions, rather than wasting time with the hardest questions and missing out on easier marks.

Read the question first

If the data looks complex, it makes sense to look at the question first before beginning to interpret the data. Just like data-heavy questions in section 1, it can take a few moments to interpret the data provided. By reading the question first, you focus your mind, giving you a better focus to approach the data with and ensuring you only spend time analysing data you actually need to work from.

Top tip! Don't spend too long on any one question. In the time it takes to answer one hard question, you could gain three times the marks by answering three easier questions. *Make the most of every second!*

Example Question

An online company provides personalised sports kit with discounts for bulk purchases. Shipping rates are £4.99. All prices are quoted in pounds and are per item.

No of items	Plain T-shirt	Polo Shirt	Long sleeve T-shirt
1-10	4.99	5.59	5.99
11-50	4.49	5.09	5.49
51+	3.99	4.49	4.99

Having read the instructions you know what the table will show. Now look straight at the question below so you know what to do with the data.

No of Items	Monotone Print	Multi-tone Print	Embroidered Logo
1-6	0.99	1.99	3.99
7-25	0.49	1.29	3.49
26+	0.29	0.89	2.99

A local hockey team requires 26 polo shirts with embroidered logo on the front and printed monotone number on the back. How much will this cost?

A. £194.01
B. £217.62
C. £222.61
D. £225.21
E. £240.81

This is a typical question. Find the right data in the table and start adding it up

Answer: C

This question highlights the need to read the question carefully as to use the correct data from the tables and is a relatively common type of question. If you look closely at the tables, you will realise that the number of items bracketed together changes between the tables. Watch out for this, or similar changes in unit, in the test. If you only skim over the tables you are in danger of missing this and will therefore get the question wrong. However, the calculation required is simple as is often the case, the question is more looking at your ability to pick out relevant data.

Price per polo shirt = 5.09 (base price) + 2.99 (embroidered logo) + 0.29 (monotone print) = £8.37
Price for 26 polo shirts as specified = 8.37 x 26 (number of items required) = £217.62
Total Price = 217.62 + 4.99 (shipping) = £222.61

Quantitative Reasoning Questions

SET 1

The country of Ecunemia has a somewhat complicated tax code. There are four states that make up Ecunemia: Asteria, Bolovia, Casova and Derivia. Each state has its own tax code, including different tax rates on different items. The table below represents the tax a **customer** has to pay when they purchase an item from a store. E.g. a £100 coat in Asteria would cost £110.

	Asteria	Bolovia	Casova	Derivia
Clothes	10%	15%	10%	10%
Food	5%	0%	10%	0%
Imports from other states	20%	5%	10%	15%

The customer must add the tax onto the advertised purchase price. In the case of an item falling into multiple categories (for example, in the case of Imported Food) the higher tax rate is paid and the lower rate is ignored.

Question 1:
A shopper visits a certain supermarket. Without tax, the shopper spends $50 on food, $30 on clothes and nothing on imported items. She spends $88 in total. Which state is this supermarket in?
A. Asteria
B. Bolovia
C. Casova
D. Derivia

Question 2:
Someone runs a supplier in Bolovia, supplying supermarkets in each state in Ecumenia. Each year they supply each state with 250 items of clothing, which the supermarket sells for $40 (including tax), and the supplier gets all of this revenue, minus the tax paid. A competitor in Asteria goes out of business, and this supplier has the opportunity to buy the manufacturing plant for $20,000, and transfer to this state.

If the supplier purchases the site, and moves to Asteria, how many years will it take to make back the cost of purchasing the site?
A. 5 years
B. 12 years
C. 23 years
D. 26 years

Question 3:
John goes into a store and spends $100. Of this, $12 is tax. Which of the following is possible?
A. He shopped in Asteria and bought no imported goods.
B. He shopped in Casova.
C. He shopped in Derivia and bought at least $50 of food (excluding tax).
D. He shopped in Bolovia and spent $10 on imported goods (excluding tax).

Question 4:
Sibella is on a road trip through Ecunemia, driving through different states. On the journey she buys $100 of the finest Asterian ham, $30 of the finest Bolovian caviar, a $10 case of Casovan orange juice and spends $100 on a Derivian dress (all of these prices without tax). Which of the following cannot have been the total amount Sibella spent, including tax?
A. $256
B. $264
C. $273
D. $288

SET 2

As a probe drops through the ocean, the pressure it experiences increases. For every 10 metres the probe drops down, the pressure it experiences increases by 10,000 Pascals (Pa).

Question 5:
A particular probe can survive 200 pounds per square inch without incurring damage. Given that the conversion factor between these units is 7000 Pa = 1 pound per square inch and assuming that pressure at sea level is 0 Pascals, how deep can the probe drop into the ocean without incurring damage?

A. 14 m
B. 140 m
C. 1.4 km
D. 14 km

Question 6:
A different probe is dropped into the ocean and falls downward. This probe can withstand 300,000 Pa of pressure without breaking. A model of the effect of the fluid states that the object's depth in the fluid is $d = \frac{1}{2}\sqrt{(t^3)}$, where d is depth in metres and t is time in seconds. How long will it take for this probe to break?

A. 65 seconds
B. 71 seconds
C. 75 seconds
D. 78 seconds.

SET 3

The fictional drug Cordrazine is used to treat four separate conditions. The following table gives the amount of drug used in each case to treat each condition, written in the form x mg/kg: i.e. for every kilogram you weigh, you take x mg of the drug. The recommended course for the drug is also listed, in the form of number of times a day and how many weeks you need to take the drug.

Condition	Dosage	Course
Black Trump Virus	4 mg/kg	3 times daily for 4 weeks
Swamp Fever	3 mg/kg	Once daily, 1 week
Yellow Tick	1 mg/kg	2 times daily for 12 weeks
Red Rage	5 mg/kg	2 times daily, 3 weeks

Question 7:

Over the course of treatment, John, an 80 kg male, takes 26.88 grams of the drug. Which disease was he prescribed the drug for?

A. Black Trump Virus

B. Swamp Fever

C. Yellow Tick

D. Red Rage

Question 8:

Carol is a 60 kg female who is prescribed the drug (precisely and at different times) three times in one year. Two of the cases are for Yellow Tick. In total she takes 40.32 grams of the drug. Which was the third disease she was prescribed the drug for?

A. Black Trump Virus

B. Swamp Fever

C. Yellow Tick

D. Red Rage

Question 9:

Clarence takes the drug twice in his life. Once he takes it for Swamp Fever at age 18, when he weighs 80 kg, and he takes it later in life at age 40 for Black Trump Virus, when he weighs 110 kg. What is the ratio of the amount he takes each time?

A. 1:23

B. 1:22

C. 1:21

D. 1:20

Question 10:

Danny has liver disease. His system cannot cope with more than 15.5 grams of Cordrazine every 4 weeks. Danny has a dental condition usually treated with Cordrazine, but dentists have advised him to not complete a course of the treatment, as he would exceed the dose that his system is able to cope with. Which of the following statements is possible?

A. Danny suffers from Red Rage and weighs 75 kg.

B. Danny suffers from Swamp Fever and weighs 100 kg.

C. Danny suffers from Black Trump and weighs 45 kg.

D. Danny suffers from Yellow Tick and weighs 75 kg.

Question 11:

Eileen has kidney failure. Her system cannot cope with more than 10 grams of Cordrazine every 4 weeks. She suffers from Red Rage, but dentists have recommended she does not use Cordrazine to treat it, as this would exceed the 10 g dosage her system can cope with. Which of the following weights is the minimum that would support this recommendation?

A. 40.34 kg

B. 42.53 kg

C. 45.81 kg

D. 47.62 kg

SET 4

A bakery sells four varieties of cakes. The cakes contain the following ingredients:

	Sponge (520g)	Madeira (825g)	Pound (710g)	Chocolate (885g)
Flour (g)	125	250	150	200
Butter (g)	125	175	185	175
Egg (g)	120	180	180	120
Milk (g)	25	45	45	150
Sugar (g)	125	175	150	200
Cocoa (g)	-	-	-	40

Question 12:
Which cake contains the highest proportion of flour?
A. Sponge
B. Madeira

C. Pound
D. Chocolate.

Question 13:
The cake recipes are scaled up for a large order. One cake weighs 2.6 kg and contains 625 g of flour. What variety of cake is it?
A. Sponge
B. Madeira

C. Pound
D. Chocolate

Question 14:
Eliza is having a wedding and wants to produce a 4-tiered wedding cake. She wishes each tier to be of different size, and scaled such that that the bottom cake is 50% heavier than normal (e.g. the cake contains 50% more ingredients), the second cake is 25% heavier than normal, the third cake is 10% heavier than normal and the top cake is normal-sized, where each cake is of the same type.

Which of the following is a possible weight of sugar for the cake (rounded to 2 s.f.)?
A. 940 g
B. 970 g

C. 1,000 g
D. 1,030 g

Question 15:
It is known that flour costs £0.55 per 1.5 kg and sugar costs £0.70 per 1 kg. Which of the following is the closest to the cost ratio of flour to sugar in a Madeira cake?
A. 1:2
B. 3:4

C. 4:5
D. 5:6

Question 16:
Milk costs £0.44 per kilogram and flour costs £0.55 per 1.5 kg. What is the cost ratio of flour to milk in a chocolate cake?

A. 1:1
B. 2:3

C. 8:7
D. 10:9

SET 5

The Kryptos Virus is particularly virulent. The infection rate is dependent upon the gender of the recipient. A random sample of 100 men and 100 women are taken from a population and tested for the Kryptos virus using Test A. The results of Test A are displayed below:

	Men	Women
Have virus	45	63
Do not have virus	55	37

Question 17:
What percentage of people tested have the virus?
A. 45%
B. 54%
C. 55%
D. 63%

Question 18:
A population of 231,768 is divided: 53% women, 47% men. Use the data in the table to estimate the number of people in the population that have the Kryptos virus. Assume that the infection rates in each gender will be the same as for the sample population in Test A. Which of the following is the number of people expected to be infected with Kryptos virus in this population?
A. 123,587
B. 123,589
C. 125,541
D. 126,406

Question: 19
3/9 of the men and 5/7 of the women testing positive for Kryptos in Test A have visited the city of Atlantis. Which of the following is the correct percentage of people in the test group testing positive for Kryptos who have **NOT** visited Atlantis?
A. 40%
B. 44%
C. 50%
D. 55%

Question 20:
It is known that Test A is not always correct. Test B is a more accurate test. The 45 men who tested positive for the Kryptos virus using Test A were then re-tested with Test B - only 20 tested positive. Assuming the same proportion of men and women experienced false positive results with Test A, how many women in the test group do we expect to actually have the Kryptos virus?
A. 20
B. 28
C. 35
D. 42

Question 21:
It is decided the women who tested positive under test A should be retested using test B. This time 29 women test positive for the Kryptos Virus. Considering both the men and women tested, what percentage of people who tested positive in Test A also tested positive in Test B (to the nearest whole number)?
A. 40%
B. 45%
C. 50%
D. 55%

For 180 more QR questions check the *Ultimate UKCAT Guide*– flick to the back to get a free copy.

Abstract Reasoning

The Basics

The abstract reasoning section of the UKCAT will test your ability to think beyond the information that is readily available to you in form of the information provided by the question. The idea behind this section of the paper is to test how well the candidate is able to respond to questions that may go beyond the scope of their knowledge or require them to apply their existing knowledge in an unusual way. This is thought to be helpful in determining how well a student will be able to interpret information such as scans, X-rays or other test results as a clinician.

This section of the test examines pattern recognition and the logical approach to a series of symbols in order to match symbols to one group or another. There are a number of different question types, but all require one key skill – the ability to recognise patterns in a set of shapes

In this section of the UKCAT, you have to answer 55 questions in only 13 minutes (with one additional minute to read instructions). Thus, it is mathematically the most time pressured section of the UKCAT. But in terms of timing, think of it in terms of the image sets. There are multiple questions per image set. Since the main investment in time is in figuring out the pattern, you have a greater proportion of the time to spend on the first question in each set. Then all subsequent questions in that set will be easy and quick to answer. By far the hardest task is deducing the rules – once you have them, matching the options to the correct set is straightforward. Therefore as a rule of thumb, if the image set has 5 questions on it you have about 60 seconds to work out the pattern. Then, match the options to the set they each belong in using the remaining time allocation.

Techniques

Timings

As with the rest of the test, you have to keep an eye on the time to keep track of how you're doing. Make sure that you stay within your time limit of 78 seconds for each block of 5 questions, and quicker for data sets with fewer questions. When divided up to 15 seconds per question it might not seem like a lot, but actually given the format of the questions you will begin to realise that it is enough. As mentioned above, for the 5-question sets keep ticking on at a steady rate of about 55–60 seconds to find the rule then about 18–23 seconds to decide which set the 5 different options fit into. Your **practice will increase both your speed and overall likelihood of finding the rule**, but despite thorough preparation you still may fail to spot the pattern. If you can't see it, don't despair – simply make reasonable guesses (you have a 33% success rate by chance alone), flag for review (in case you have spare time at the end to check back and have another go) and move swiftly on.

Top tip! Give yourself plenty of time to systematically work out the rules. Once you have found them, answering the questions will be quick.

Pattern Recognition

By far the most important ability in this section is to correctly identify patterns, as regardless of question style the matching process is straightforward once you have identified the rules. Some people are naturally better at this than others. You might be the sort of person who sees these patterns easily and can quickly put a name to the rule, or you might be the sort of person who finds it takes them more time and effort to work out what's going on. In reality, everyone lies at a different position on this scale, but one thing is certain. **You can improve your speed and accuracy on this section by having a methodical system** that can be repeated and applied to all shape-sets. One such system is the **NSPCC** system. This provides a logical structure for working through each set of images and looking for different components of a possible pattern.

In this system, the letters stand for:

<u>N</u>umber → <u>S</u>ize → <u>P</u>osition → <u>C</u>olour → <u>C</u>onformation

Using this system, you consider each of the following aspects of the images in sequence, looking each time for commonly used patterns. We recommend this because it begins by looking for the simplest and most common potential patterns – if they are present, you are sure to get the pattern quickly and easily. If the first few patterns you look for are not present, then you look further on in the sequence to check for harder and less commonly tested patterns until you arrive at the answer.

Practice

It's very important to practice for this section as the style of questions are unlikely to be familiar. **Practicing well gives you three key advantages**. Firstly, you get used to the types of patterns which are likely to be asked in the real exam. This makes it more likely you will spot the patterns quickly as you will have seen them before, and it also trains up your implicit recognition system, meaning that if you take an "educated guess" you are more likely to be right. Secondly, it gives you practice implementing a pattern recognition system, like **NSPCC**. With practice you will become better at using the system, and therefore quicker and more accurate overall. Thirdly, you will gain a feeling for the time it takes to answer different types of question. This will allow you to better plan your time on the day, making the most out of every second you have.

Guessing

If you practice well you shouldn't have to guess very many questions, but it might be necessary if you just can't figure out a pattern. Guessing in this section of the test has a reasonable probability of success. Since there are only three or four options per question there is a 25-33% chance of guessing any one question correctly, and if all 5 questions in a data set are guessed then there is and 87% chance of gaining at least one mark from the set – better than in any other section of the UKCAT. However there is more to it than that.

Whilst the best way to answer these questions correctly is to formally deduce the rule and apply it (using a pattern recognition system like **NSPCC** helps here), humans *do* have an innate instinct for pattern recognition. This innate instinct is not necessarily right and can lead you astray, but in a quick guessing situation, it can be applied cleverly to boost your chances of guessing correctly. That is to say, in some questions the overall look of the image will feel as though it should be placed in a particular set – you wouldn't be able to say exactly why, but to you it would look much more like one group than the other. Learning to harness this power can help give you a much better guessing accuracy. Below is a simple example to demonstrate:

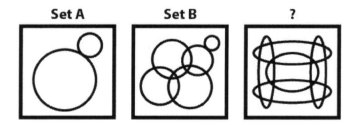

By quickly looking at the images provided for Set A and B, you get a general feel for what the contents are like. If asked which set the question image belongs to, you might be tempted to say Set B – it kind of looks more complicated and cluttered then Set A. It just looks more like the image. Now whilst this reasoning does not provide a comprehensive rule, in instances like this it can lead you to the correct answer quickly – even if you don't properly identify the underlying pattern.

How to use the NSPCC system

The **NSPCC** system is a good way of working through possible patterns in a structured way. Whilst no structured response can be perfect, this system will solve over 90% of patterns quicker and more reliably than by trusting intuition alone.

Using the system, it is important that you are thorough. Sometimes the pattern can be subtle and you could easily miss out on it if not taking care. You have to examine the details closely: count corners, sides of shapes, check where they are etc. **You shouldn't be looking AT the shapes; you should be looking FOR patterns**. You should be working quickly, checking one thing, and if it's not that checking the next item in your list until you find the rule. Now to look at the system step by step.....

Number

Looking at "number" is about *counting* as many things as possible. How many dots? How many squares? How many sides? How many corners? How many right angles? Also have a look at how many different types of shapes you can find in the frames. Sometimes you might find one type of shape only in Set A and not in Set B for example. A good rule of thumb: block arrows have 7 sides, so don't count all the sides individually every time you see one!

Size

It is quite common to find patterns in the size of the shapes. Is one shape always bigger than the others? Is there always a big shape in the centre, or in the corner? Are there smaller shapes inside larger ones?

Position

Look for patterns in where shapes are positioned. You might, for instance, always find a square in the top right corner in one set and a circle in the top right corner of the other set. Look for Look also for touching and overlap of shapes – when you see this, make careful note of the type of contact. Is it tangential? Does it cut the shape in equal pieces, or is it off centre? Is there a certain shape that always makes this contact?

Colour

The shading of different shapes can constitute a pattern. Whilst this is often the easiest pattern to spot, it takes its place lower down in the system as it can often be a distracter. Most diagrams contain some amount of shading, but only occasionally is the primary pattern centred upon this. Look for shapes that are always shaded. Are all triangles black in one set and all circles black in the other, for example? On the other hand, are some shapes never shaded?

Conformation

These are the hardest patterns to spot, as they are the more complex patterns that can't be found by looking at the more geometric aspects. Conformation describes the pattern by which the shapes are arranged within the box – so you have to take a step back and look at the box as a whole in order to spot them. Look for patterns to the arrangement, like shapes arranged in a horizontal, vertical or diagonal line. Look also for the influence of one shape on another. For example, the presence of a white circle might signal a 90 degrees clockwise rotation of one shape and the presence of a black circle might signal a 90 degrees anticlockwise rotation, for example. When there are arrows, look at where they are pointing: are they all pointing in the same direction or at the same thing? You're looking for second order patterns, how things change based upon other aspects of the image.

Question Answering Strategy

There are four styles of question in this section, but all of these require the same pattern-recognition skills.

The **first style** of questions is the original style of question that used to be the only style in this section of the UKCAT. It also tends to be the style that accounts for the majority of the questions in this section, however this is not an official rule and it will not necessarily be the same this year. In this type of question, you are provided with two sets of six shapes, Set A and Set B. All of the images within each set are linked to each other by a common rule, but the rule must be different for Set A and Set B. The task is to identify the rules for each set, then for the 5 options you need to decide where they belong. If an option follows the same rule as the shapes in Set A, then it belongs in Set A, and likewise for Set B. If the image obeys neither the rule for Set A nor the rule for Set B, then it is correct to say it belongs in neither set and you choose the "neither" option. Approach this style of question by spending the majority of the time deducing the rule by using a system like **NSPCC**. Then when you have decided on the rule for each set, work through the options, selecting which set each fits best. If you don't figure out the rule in time – don't worry. It is expected that you won't work out them all in the time pressure of the exam. In that case, simply use an educated guessing strategy to give yourself a good chance of picking up some marks and then move on.

> *Top tip!* If a shape fits the rules for **both sets**, then the correct response is always "**neither**"

In the **second style** of question, you are provided with a single sequence of four shapes. They should be read as a sequence from left to right. You are then asked to choose the next shape in the sequence out of four options. This style of question is normally quicker to answer as it is more intuitive. In addition, you have three different transition points that you can compare to each other to help deduce whether you have correctly identified the rule. To answer this style of question well, start by scanning quickly across all four shapes – this gives you a general understanding of what is happening in the sequence. Then, focus on the element that is changing and apply your system to find out exactly *how* it is changing. Once again, if you're struggling, you can probably use your intuition to make a decent educated guess and move on. Consider flagging for review so you know where to focus your efforts if you have any additional time at the end.

The **third style** of question is also a sequence style of question. You are provided with two shapes with a rule linking them. The rubric states shape one **is to** shape two by this rule. Then you are provided with shape three, and you have to apply this rule to find out how the rule transforms shape three into shape four (you are given four options). Once again, the key to accuracy is in deducing the rule that links the shapes. Focus on the same elements in shape one and shape two and notice any changes that have taken place. This style of question has a more straightforward strategy as there are many fewer options to examine; as you can directly compare the two boxes you can usually deduce the rule without needing to use a system, but remember it is always there if needed. Be certain to check the rule applies to *all* elements in the boxes, otherwise you will need to revise the rule to account for the box as a whole rather than just one or two of the elements. Once you are satisfied with the change, apply the rule to the next shape and select the answer. Ideally you should imagine what the answer is *before* focusing on the options, otherwise you could be biased by a similar but incorrect option, but if you are struggling to do this then use your intuition to select the option that feels as though it is the best fit.

The **fourth style** of question is very similar to the first and original style. You are provided with two sets of six shapes, Set A and Set B, each linked by a common rule. Once again apply the same system to deduce the rule for each set. The only difference comes when it is time to select your answer. Instead of being asked which set a shape fits into, you are provided with four options and asked to select the one that fits into one set or the other. So once you know the rule, test the options to see which one fits the set you are asked for.

Example 1

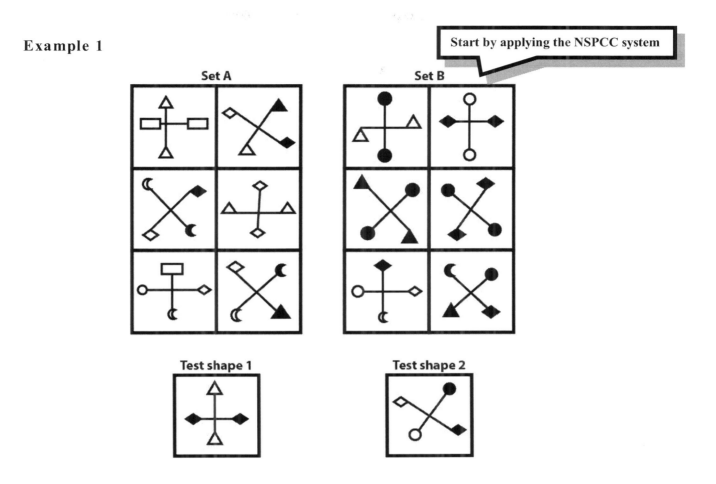

Start by applying the NSPCC system. Number: count the number of shapes, sides, angles and so on looking for a pattern (I can't see one). Size: are all shapes the same size (yes they are). Position: is there a pattern to where certain shapes are (not obviously). Colour: Is there a pattern to the shading (yes, the shading is dependent upon the shape of the cross). If you didn't get that, look back now to identify what the pattern is before reading on.

We can use our observations to devise the following rules:
Set A: (+)-shaped crosses have four white shapes and (X)-shaped crosses have two white and two black shapes.
Set B: (+)-shaped crosses have two white and two black shapes and (X)-shaped crosses have four black shapes.

Applying these rules tells us Test Shape 1 belongs to Set B and Test Shape 2 belongs to Set A.

A Final Word

This section is all about pattern recognition. The more you see the better you will become. Once you're familiar with the main types of patterns which come up, you'll be able to solve the majority of questions without difficulty. Remember that **you're looking to identify a rule** for each set of boxes, something which links them all together. Then, you can decide which set each question item fits into (or indeed neither). Start using the **NSPCC** system, then practice makes perfect!

Abstract Reasoning Questions

SET 1

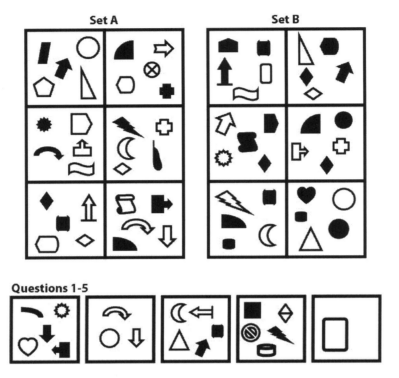

SET 2

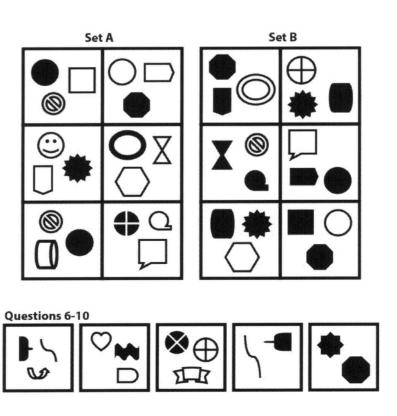

SET 3

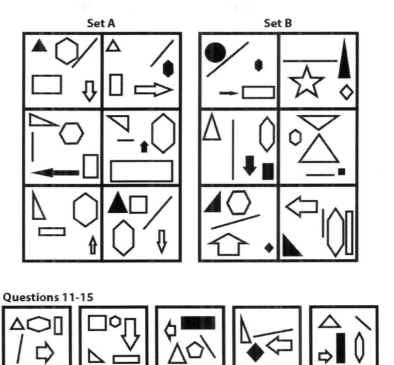

Questions 11-15

SET 4

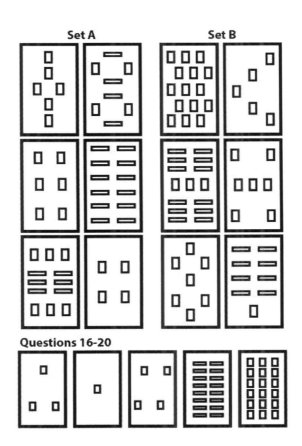

Questions 16-20

SET 5

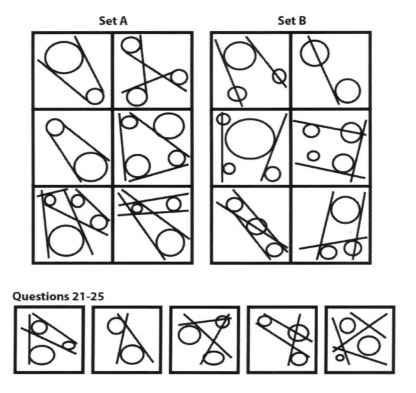

Questions 21-25

SET 6

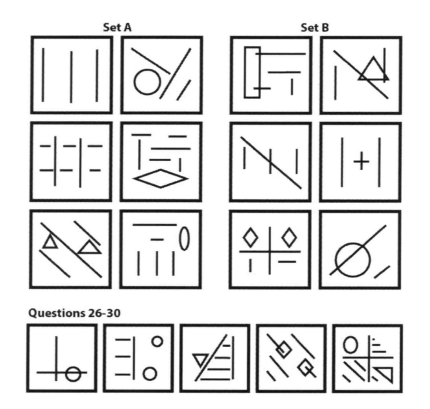

Questions 26-30

SET 7

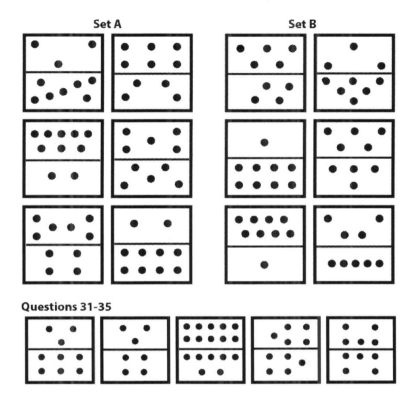

Questions 31-35

SET 8

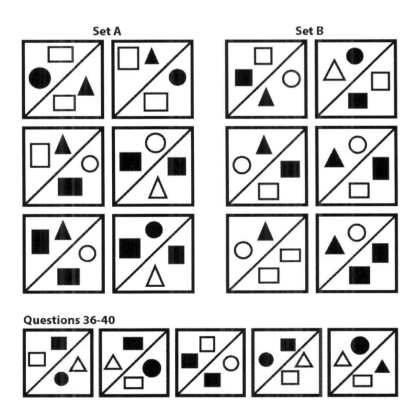

Questions 36-40

For Questions 41 – 45: which answer option completes the series?

Q41

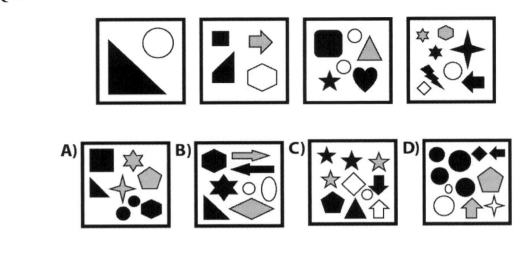

Q42

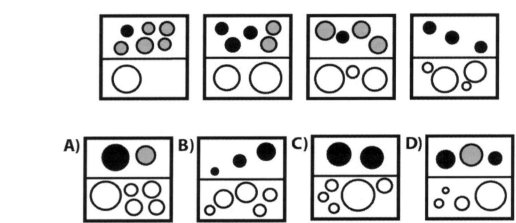

Q43

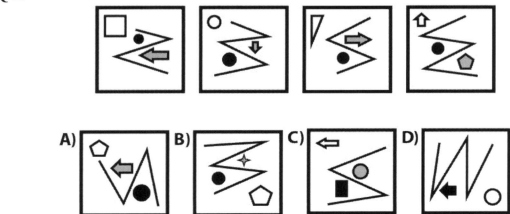

Q44

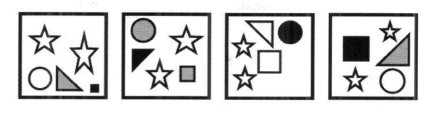

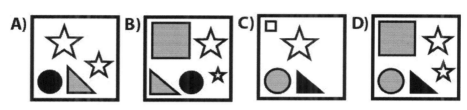

Q45

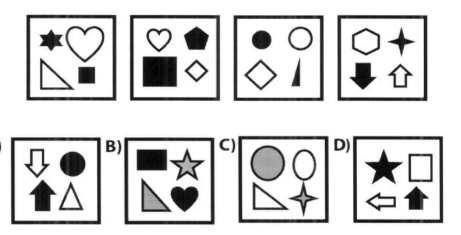

For 150 more AR questions check the *Ultimate UKCAT Guide*– flick to the back to get a free copy.

Situational Judgement

The Basics

Situational judgment is a psychological aptitude test; it is an assessment method used to evaluate your ability in solving problems in work-related situations. Situational Judgement Tests (or SJTs) are widely used in dentistry as one of the criteria when deciding on applicants; it is used for the Dental Foundation Programme and Dental Core Training applications.

The aim of the situational judgment section is to assess your ability to understand situations that you could encounter as a dental student or dentist and how you would deal with them. (The UKCAT is the same for medical and dental school applicants so many questions may be medically based, however you do not need any medical knowledge) It is a method to **test some of the qualities required in a healthcare professional** (e.g. integrity and ability to work in a team). As this test is for dental and medical student, some of the topics may not be solely aimed at dental students.

In the UKCAT, the situational judgment section consists of 20 scenarios with 67 items. Each scenario will have 3-6 items. You will have 26 Minutes to complete this section, which translates to approximately 22 seconds per item. As with all UKCAT sections, you have one additional minute to read the instructions at the start. Ensure you're careful to mark your intended answers when working at this pace.

This is the last section of your exam, you are almost near the end- only twenty seven more minutes to go. You still need to stay focused; it might seem obvious, but make sure you read the whole scenario and understand it prior to answering. When answering, imagine you are the person in the scenario. The majority of the scenarios will be about medical students, imagine you are in their shoes.

The series of scenarios include possible actions and considerations. Each scenario is comprised of two sets of questions. In set one you will be asked to assess the "appropriateness" of options in relation to the scenario.

The four possible <u>appropriateness</u> choices are:

➤ *A very appropriate thing to do* – This is an ideal action.
➤ *Appropriate, but not ideal* – This option can be done but not necessarily the best thing to be done.
➤ *Inappropriate, but not awful* – This should not be done, but if it does occur the consequences are not terrible.
➤ *A very inappropriate thing to do* – This should not be done in any circumstances, as it will make the situation worse.

In set two, you will be asked to assess the "importance" of options in relation to the scenario,

The four possible <u>importance</u> options are:

➤ *Very important* – something that is essential to take into account.
➤ *Important* – something you should take into account but is not vital.
➤ *Of minor importance* – something that may be considered, but will not affect the outcome if it is not taken into account.
➤ *Not important at all* – something that is not relevant at all.

Section E includes non-cognitive abilities and so is marked differently to the other sections. You will be awarded full marks if your response completely matches the correct answer. If your answer is close but not exactly right you will receive partial marks, and if there are no correct answers you will receive no marks. Your score is then calculated and expressed in one of four bands, band 1 being the highest and band 4 being the lowest. The band resembles how close your responses were to the assessment panel's agreed answers.

Things you MUST NOT do as a dental student:

➢ Sign or authorise anything that a qualified dentist should do (prescriptions etc).
➢ Make any decisions that affect treatment without consulting the tutors.
➢ Change any treatment regime for any reason made by consultants.
➢ Perform practical procedures without supervision.
➢ Break patient confidentiality or do anything that places confidentiality at risk.
➢ Behave dishonestly in any way.

Things you MUST do as a dental student:

➢ Raise any concerns regarding patient safety with an appropriate person.
➢ Report any inappropriate behaviour you witness to an appropriate person.
➢ Attend all scheduled teaching and training.
➢ Take responsibility for your learning and seek opportunities to learn.
➢ Dress and present yourself in a clean and smart way.
➢ Never put patients at risk or harm

Things you CAN do as a dental student:

➢ Speak to patients.
➢ Examine patients.
➢ Write in patient notes (but must indicate your name and role).
➢ Perform simple practical procedures like restorations, root canal treatment, crown preparations (if you are trained to and with adequate supervision).
➢ Assist dentists with more complex procedures according to their strict directions.
➢ Attend meetings where patients are discussed.
➢ Do things you have been trained to do.

> *Top tip!* Always put patient safety first. If you do this, you can never go far wrong.

Question Answering Strategy

Treat every option as independent – the options may seem similar, but don't let the different options confuse you, read each option as if it is a question on its own. It is important to know that responses should **NOT** be judged as though they are the **ONLY** thing you are to do. An answer should not be judged as inappropriate because it incomplete, but only if there is some actual inappropriate action taking place. For example, if a scenario says "a patient is in the clinic and complains she is in pain", the response "ask the patient what is causing the problem" would be very appropriate, even though any good response would also include informing the nurses and dentists about what you had been told.

There might be multiple correct responses for each scenario, so don't feel you have to answer each stem differently. Thus an answer choice may be used once, more than once or not at all for all scenarios.

If you are unsure of the answer, mark the question and move on. Avoid spending longer than 30 seconds on any question, otherwise you will fall behind the pace and not finish the section.

As with every other section, if you are completely unsure of the answers, answer the question anyway. There is no negative marking and your initial instinct may be close to the intended answer.

➢ If there are several people mentioned in the scenario make sure you are answering about the correct person.
➢ Think of what you 'should' do rather than what you necessarily would do.
➢ Always think of **patient safety** and acting in the patient's best interests.

Read "GDC Standards"

This is a publication produced by the GDC (General Dental Council) which can be found on their website. The GDC regulate the dental profession, ensuring standards remain high. This publication can be found on their website, and it outlines the standards that dentists are judged against and are expected to follow. **Reading through this will get you into a professional way of thinking** that will help you judge these questions accurately.

Step into Character

When doing this section, imagine you're there. Imagine yourself as a caring and conscientious dental student a few years from now, in each situation as it unfolds. What would you do? What do you think would be the right thing to do?

Hierarchy

The patient is of primary importance. All decisions that affect patient care should be made to benefit the patient. Of secondary importance are your work colleagues. So if there is no risk to patients, you should help out your colleagues and avoid doing anything that would undermine them or harm their reputation – but if doing so would bring detriment to any patient then the patients priorities come to the top. Finally of lowest importance is yourself. You should avoid working outside hours and strive to further your education, but not at the expense of more important or urgent priorities. Remember the key principles of professional conduct and you cannot go far wrong. **Of first and foremost importance is patient safety**. Make sure you make all judgements with this in mind.

> *Top tip!* Read the GDC publication "*GDC Standards*" – this will help you think the right way.

Dental Ethics

There tend to be a few ethical questions in each SJT paper so it is well worth your time to learn dental ethics. Whilst there are huge ethical textbooks available– you only need to be familiar with the basic principles for the purposes of the UKCAT. These principles can be applied to all cases regardless what the social/ethnic background the healthcare professional or patient is from. In addition to being helpful in the UKCAT, you'll need to know them for the interview stages anyway so they're well worth learning now rather than later. The principles are:

Beneficence

The wellbeing of the patient should be the dentist's first priority. In dentistry this means that one must act in the patient's best interests to ensure the best outcome is achieved for them i.e. 'Do Good'.

Non-Maleficence

This is the principle of avoiding harm to the patient (i.e. Do no harm). There can be a danger that in a willingness to treat, dentists can sometimes cause more harm to the patient than good. This can especially be the case with major interventions, such surgery. Where a course of action has both potential harms and potential benefits, non-maleficence must be balanced against beneficence.

Autonomy

The patient has the right to determine their own health care. This therefore requires the dentist to be a good communicator, so that the patient is sufficiently informed to make their own decisions. 'Informed consent' is thus a vital precursor to any treatment. A dentist must respect a patient's refusal for treatment even if they think it is not the correct choice. Note that patients cannot <u>demand</u> treatment – only refuse it, e.g. an alcoholic patient can refuse rehabilitation but cannot demand a liver transplant.

There are many situations where the application of autonomy can be quite complex, for example:

➢ **Treating Children**: Consent is required from the parents, although the autonomy of the child is increasingly taken into account as they get older.

➢ **Treating adults without the capacity** to make important decisions. The first challenge with this is in assessing whether or not a patient has the capacity to make the decisions. Just because a patient has a mental illness does not necessarily mean that they lack the capacity to make decisions about their health care. Where patients do lack capacity, the power to make decisions is transferred to the next of kin (or Legal Power of Attorney, if one has been set up).

Justice

This principle deals with the fair distribution and allocation of healthcare resources for the population.

Consent

This is an extension of Autonomy – patients must agree to a procedure or intervention. For consent to be valid, it must be **voluntary informed consent.** This means that the patient must have sufficient mental capacity to make the decision, they must be presented with all the relevant information (benefits, side effects and the likely complications) in a way they can understand and they must make the choice freely without being put under pressure.

> *Top tip!* Remember that **consent is only valid** if it is given:
>
> ➢ On the basis of full information.
> ➢ With sufficient mental capacity.
> ➢ Freely without pressure.
> ➢ Communicated unambiguously.

Confidentiality

Patients expect that the information they reveal to dentists will be kept private- this is a key component in maintaining the trust between patients and dentists. You must ensure that patient details are kept confidential. Confidentiality can be broken if you suspect that a patient is a risk to themselves or to others e.g. Terrorism, suicides.

When answering a question on dental ethics, you need to ensure that you show an appreciation for the fact that there are often two sides to the argument. Where appropriate, you should outline both points of view and how they pertain to the main principles of dental ethics and then come to a reasoned judgement.

Situational Judgement Questions

Scenario 1

A conversation is taking place between a midwife Kate and the senior Dr Herbert: Jacob, the medical student, is observing. Dr Herbert is being rude to the Kate and is acting superior. When Dr Herbert leaves, Jacob overhears Kate talking to the other midwives about his behaviour, and how it happens frequently, and makes both the midwives and the patients feel uncomfortable.

How <u>appropriate</u> are the following actions from <u>Jacob</u>?

1. Tell Kate that you will help to file a complaint against Dr Herbert.
2. Make Dr Herbert aware that perhaps he should be kinder the next time he speaks to Kate and patients.
3. Ignore the situation and do nothing.
4. Alert his supervisor as to what he saw, and to get advice on what to do.
5. Tell Dr Herbert that his behaviour was making patients and midwives feel uncomfortable.

Scenario 2

A medical student, George, is sitting in a foot clinic with Dr Walker. George notices that Dr Walker hasn't been washing his hands between patients, despite examining the feet of all of his patients without gloves. In his training George was told that he must wash his hands properly before and after touching each patient to prevent the spread of infections.

How <u>appropriate</u> are each of the following responses by <u>George</u> in this situation?

6. Alert Dr Walker that he ought to wash his hands more after the current consultation has finished.
7. Wash his hands before and after each patient in the hopes that Dr Walker will follow by example.
8. Do nothing because Dr Walker is an experienced consultant.
9. Tell the nurse in charge of the foot patients after the clinic has finished.
10. Write in the patient notes that Dr Walker didn't wash his hands before examining them.

Scenario 3

A medical student, Linh, is working on a project with a small group of other students. The students have to examine real skull bones, which were provided by the medical school's museum, and are very valuable. One of the students in Linh's group accidentally drops the skull and some of the smaller delicate bones shatter.

How <u>appropriate</u> are the following responses by <u>Linh</u>?

11. Ignore what happened, throw the skull remains away, and borrow another group's skull to finish the project.
12. Alert the museum curator about what happened as a group, and write a letter of apology together.
13. Pretend that the skull was stolen.
14. Tell the museum curator in private about who dropped the skull.
15. Tell her supervisor.

Scenario 4

A student, Henry, is living in a set of halls with students that study many different subjects. The other students find it funny to joke about Henry's work. Henry is finding it difficult to keep up with his work, and silently takes offense every time the other students joke with him. The night before one of Henry's exams, the other students make a joke that really affects Henry, and he is unable to concentrate on finishing up his revision.

How <u>appropriate</u> are each of the following responses by <u>Henry</u>?

16. Speak to his personal tutor about how he can organise himself and tackle his work in the future.
17. Retaliate by insulting the other students.
18. Do nothing because he doesn't want to offend anyone and is embarrassed about not being able to cope with the workload.
19. Move out of the halls.
20. Speak to his friends about how annoying he finds his flat mates.

Scenario 5

Mark, a medical student, is working with a group of nursing and physiotherapy students to learn about integrated care. Mark is mistaken for a junior doctor, as he is not in uniform, and is asked to test the urine of an elderly patient on the ward using a dipstick. Mark is familiar with the patient, and knows exactly how to do the test. Unfortunately, the doctor that asked him to do the test has left, and there are no other members of staff that are able to do the test for another 5 hours. The results of the test will determine the patient's management.

How <u>appropriate</u> are the following responses by <u>Mark</u>?

21. Get the most senior student in his study group to perform the test and write the results in the patient's notes.
22. Do the test himself and write the results in the patient's notes.
23. Bleep the den doctor that is in charge of the patient to alert him about his mistake.
24. Pretend that the doctor never asked him to do the test.
25. Try to find another member of staff that would be capable of performing the test.

Scenario 6

A dental student, Adele, is studying for her first year exams. She has started to panic, and does not feel as though she will be able to complete her revision before the exams start. If Adele fails the exams she would have to resit them in her holidays, which she has come to terms with. She is embarrassed of the possibility of failing, and would rather tell her friends and family that she was ill and unable to take the exams than face the embarrassment of failure. It is against the dental schools rules to opt out of an exam without a medical reason and a Doctors letter.

How <u>appropriate</u> are the following actions for <u>Adele</u> to take?

26. Fake an illness and postpone her exams.
27. Speak to her parents and her personal tutor about her struggle to get through the revision.
28. Speak to the other dental students to see if they all felt the same way about their work.
29. Refuse to turn up to the exams on the day and pretend that she had food poisoning.
30. Make an efficient revision plan for her remaining days before the exams and attempt to do the exams as best as she can.

Scenario 7

Daniel and Sean are dental students who are working together on a project. They get into a heated argument in the dental hospital lobby because Daniel has been prioritising his social life recently which is frustrating Sean.

How <u>important</u> are the following factors for <u>Sean</u> in deciding on what to do?

31. Sean can generally produce better work than Daniel anyway.
32. The mark that they get will be recorded in their log books.
33. Daniel and Sean have to work together for the rest of the year.
34. Daniel has recently broken up with his girlfriend.
35. Sean usually does most of the work when they have to do projects together.

Scenario 8

A medical student Tanya is invited to attend a clinic with Dr Garg who is in charge of Tanya's grade for the term. On the morning of the clinic, Tanya realises that she has not finished her essay that is due the next day.

How <u>important</u> are the following factors for <u>Tanya</u> to consider in deciding on what to do?

36. The importance of the essay towards her final mark for the year.
37. Tanya's friend did not find the clinic very educational.
38. Tanya's reputation with Dr Garg.
39. Whether or not Tanya will be able to attend a different clinic with Dr Garg.
40. How long it will take to finish the essay.

For 200 more SJT questions check the *Ultimate UKCAT Guide*– flick to the back to get a free copy.

UKCAT ANSWERS

Verbal Reasoning Answers

Q	A	Q	A	Q	A	Q	A
1	False	11	False	21	C	31	C
2	False	12	False	22	B	32	D
3	True	13	Can't tell	23	D	33	B
4	Can't tell	14	False	24	C	34	A
5	Can't tell	15	Can't tell	25	D	35	D
6	False	16	False	26	A		
7	False	17	Can't tell	27	C		
8	Can't tell	18	True	28	C		
9	Can't tell	19	Can't tell	29	A		
10	True	20	True	30	C		

Set 1:

1. **False** -In paragraph 2, the passage says that only the countries that signed the protocol were legally bound.
2. **False** -In 2004, the condition that over 55% of emissions were accounted for was met without Australia and the US.
3. **True** -in paragraph 2: 'each country that signed the protocol agreed to reduce their emissions to their own specific target'.
4. **Can't tell** -We know this is when the Kyoto Protocol was enforced but there is no information to suggest whether emissions actually decreased.
5. **Can't tell** -Paragraph 4 says that a 60% emission reduction would have a 'significant impact', but we cannot tell if the remaining effects would be harmful or not. The passage is also talking about a global reduction of 60% - there is no mention of each individual country needing reduce their emissions by 60%.

Set 2:

6. **False** -From paragraph 1, we can see that the Soviets were mainly concerned with showing off their economic power and technological superiority.
7. **False** -Paragraph 3 says that Project Apollo was tasked with landing the first man on the moon.
8. **Can't tell** -The passage does not tell us why the Soviets did not land a man on the moon.
9. **Can't tell** -We do not know the state of the American space efforts prior to the launch of Sputnik – we just know that its launch 'prompted urgency'.
10. **True** -The Soviets had just sent Yuri Gagarin into space and it wasn't expected that the US would beat the Soviets in landing a man on the moon; therefore the US was behind the Soviets at this stage.

Set 3:

11. **False**- Paragraph 1 says Pheidippides was the fastest runner
12. **False**-Paragraph 2 says that they were standardised from 1921 but using the 1908 distance
13. **Can't tell**-Paragraph 1 says the Greeks were not expecting to beat the Persians, but we do not know if the Persians were expecting to beat the Greeks, despite their larger army
14. **False**- Paragraph 2: to be an IAAF marathon the distance must be 26.2 miles. The original route was 25 miles
15. **Can't tell** -We know that the Persian soldiers outnumbered the Greeks and had superior cavalry, but we are not told about their training

Set 4:

16. **False-** From paragraph 1, we can see that birds would not survive if they did not migrate
17. **Can't tell-**Whilst there is only mention here of migrating in flocks, nowhere does it specifically state that either all migrating birds do so or that any migrating birds do not
18. **True-** Although the leading bird does not benefit at the time, they continually change position within the flock according to paragraph 3
19. **Can't tell** -Paragraph 3 shows that the Northern bald ibis does not behave in this way so it is not known what would happen
20. **True** -Paragraph 1 states they migrate north in the spring, and south in the winter

Set 5:

21. **C-**Though sandstone is made from sand, the passage does not state that ALL rocks are made from this material.
22. **B-**The passage discusses the valley when implementing the wall-building analogy.
23. **D-** 'Some ancient source' is all we are told, and so the source is undisclosed (Paragraph 3).
24. **C-** This rock is said in paragraph 2 to be found 'everywhere', so not 'nowhere'.
25. **D-** Paragraph 3: The grains are said to be sorted in groups 'of a size', i.e. measurements. They are all described as 'worn and rounded', and no mention is made of differentiation through age/shape.

Set 6:

26. **A-** Though the flowers smell pleasant, no mention is made of this scent being used to manufacture perfumes.
27. **C-** The passage states that the flower could bloom more than once a century, precluding 'D', but that it is thought to only do so in a century, providing evidence for 'C'.
28. **C-** The statement claims that Narcissus plants are 'prized by many' over Lilies, but this does not mean that all people - or even the majority of people - think the former is more attractive/better than the latter, or that all homeowners enjoy the plant.
29. **A-** It is actually a substance 'very similar to rum', not rum itself.
30. **C-** The genus 'belongs' to the family, as stated in paragraph 1.

Set 7:

31. **C-**Women in Illinois, not across USA, were subject to the law, and the passage does not state either a change in fashion or actual arrests, only the potential for arrests.
32. **D-** The pulling out of feathers from live birds was seen as the negative to using osprey feathers.
33. **B-** They could be possessed only 'in their proper season'.
34. **A-** The problem cited is that the article was already in use in the clothing of numerous military men. The authority of the princess/sexist politics does not feature in the passage, and 'D' is patently false
35. **D-** None of those are precluded, as only 'harmless' and 'dead' birds (in their entirety) were prohibited. Wearing a living bird was not explicitly banned.

Decision Making Answers

Question	Answer	Question	Answer	Question	Answer	Question	Answer
1	D	6	C	11	C	16	C
2	B	7	C	12	B	17	A
3	B	8	B	13	B	18	D
4	B & C	9	C & E	14	B	19	E
5	D	10	B	15	C	20	C

Question 1: D

The easiest thing to do is draw the relative positions. We know Harrington is north of Westside and Pilbury. We know that Twotown is between Pilbury and Westside. Crewville is south of Twotown, Westside and Harrington but we do not know but its location relative to Pilbury.

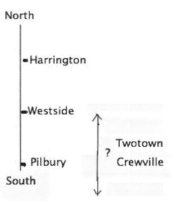

Question 2: B

By making a grid and filling in the relevant information the days Dr James works can be deduced:

	Sunday	Monday	Tuesday	Wednesday	Thursday	Friday	Saturday
Dr Evans	X	√	X	X	√	√	√
Dr James	X	√	√	√	√	X	√
Dr Luca	X	X	√	√	X	√	√

- No one works Sunday.
- All work Saturday.
- Dr Evans works Mondays and Fridays.
- Dr Luca cannot work Monday or Thursday.
- So, Dr James works Monday.
- And, Dr Evans and Dr James must work Thursday.
- Dr Evans cannot work 4 days consecutively so he cannot work Wednesday.
- Which means Dr James and Luca must work Wednesday.
- (mentioned earlier in the question) Dr Evans only works 4 days, so cannot work Tuesday.
- Which means Dr James and Luca work Tuesday.
- Dr James cannot work 5 days consecutively so cannot work Friday.
- Which means Dr Luca must work Friday.

Question 3: B

All thieves are criminals. So the circle must be fully inside the square, we are told judges cannot be criminals so the star must be completely separate from the other two.

Question 4: B and C

Using the information to make a diagram:

Hence **A** is incorrect. **D** and **E** may be true but we do not have enough information to say for sure. **B** is correct as we know peaches are more expensive than oranges but not about their price relative to pears. Equally we know **C** to be true as grapes are more expensive than apples so they must be more expensive than pears.

Question 5: D

Using Bella's statements, as she must contradicted herself with her two statements, as one of them must be true, we know that it was definitely either Charlotte or Edward. Looking to the other statements, e.g. Darcy's we know that it was either Charlotte or Bella, as only one of the two statements saying it was both of them can have been a lie. Hence it must have been Charlotte.

Question 6: C

Work through each statement and the true figures.
A. Overlap of pain and flu-like symptoms must be at least 4% (56+48-100). 4% of 150: 0.04 x 150=6
B. 30% high blood pressure and 20% diabetes, so max percentage with both must be 20%. 20% of 150: 0.2*150 = 30
C. Total number of patients – patients with flu-like symptoms – patients with high blood pressure. Assume different populations to get max number without either. 150 – (0.56 x 150) – (0.3 x 150) = 21
D. This is an obvious trap that you might fall into if you added up the percentages and noted that the total was >100%. However, this isn't a problem as patients can discussed two problems.

Question 7: C

We know that Charles is born in 2002, therefore in 2010 he must be 8. There are 3 years between Charles and Adam, and Charles is the middle grandchild. As Bertie is older than Adam, Adam must be younger than Charles so Adam must be 5 in 2010. In 2010, if Adam is 5, Bertie must be 10 (states he is double the age of Adam). The question asks for ages in 2015: Adam = 10, Bertie = 15, Charles = 13

Question 8: B

In this question it is worth remembering it will take more people a shorter amount of time.
Work out how many man hours it takes to build the house. Days x hours x builders
12 x 7 x 4 = 336 hours
Work out how many hours it will take the 7man workforce: 336/7 = 48 hours
Convert to 8 hour days: 48/8 = 6 days

Question 9: C & E

The easiest way to work this out is using a table. With the information we know:

1st		Madeira
2nd		
3rd	Jaya	
4th		

Ellen made carrot cake and it was not last. It now cannot be 1st or 3rd as these places are taken so it must be second:

1st		Madeira
2nd	Ellen	Carrot cake
3rd	Jaya	
4th		

Aleena's was better than the tiramisu, so she can't have come last, therefore Aleena must have placed first

1st	Aleena	Madeira
2nd	Ellen	Carrot cake
3rd	Jaya	
4th		

And the girl who made the Victoria sponge was better than Veronica:

1st	Aleena	Madeira
2nd	Ellen	Carrot cake
3rd	Jaya	Victoria Sponge
4th	Veronica	Tiramisu

Question 10: B

After the first round; he knocks off 8 bottles to leave 8 left on the shelf. He then puts back 4 bottles. There are therefore 12 left on the shelf. After the second round, he has hit 3 bottles and damages 6 bottles in total, and an additional 2 at the end. He then puts up 2 new bottles to leave 12 – 8 + 2 = 6 bottles left on the shelf. After the final round, John knocks off 3 bottles from the shelf to leave 3 bottles standing.

Question 11: C

Remember that pick up and drop off stops may be the same stop, therefore the minimum number of stops the bus had to make was 7. This would take 7 x 1.5 = 10.5 minutes.
Therefore the total journey time = 24 + 10.5 = 34.5 minutes.

Question 12: B

The time could be 21:25, if first 2 digits were reversed by the glass of water (21 would be reversed to give 15). **A** cannot be the answer, because this would involve altering the last 2 digits, and we can see that 25 on a digital clock, when reversed simply gives 25 (the 2 on the left becomes a 5 on the right, and the 5 on the right becomes a 2 on the left). **C** cannot be the answer, as this involves reversing the middle 2 digits. As with the right two digits, the middle 2 digits of 2:5 would simply reverse to give itself, 2:5. **D** could be the time if the 2nd and 4th digits were reversed, as they would both become 2's. However, the question says that 2 *adjacent* digits are reversed, meaning that the 2nd and 4th digits cannot be reversed as required here. **E** is not possible as it would require all four numbers to be reversed. Thus, the answer is **B**.

Question 13: B

To answer this, we simply calculate how much total room in the pan will be taken up by the food for each guest:
-    2 rashers of bacon, giving a total of 14% of the available space.
-    4 sausages, taking up a total of 12% of the available space.
-    1 egg takes up 12% of the available space.
Adding these figures together, we see that each guest's food takes up a total of 38% of the available space.
Thus, Ryan can only cook for 2 guests at once, since 38% multiplied by 3 is 114%, and we cannot use up more than 100% of the available space in the pan.

Question 14: B
The trains come into the station together every 40 minutes, as the lowest common multiple of 2, 5 and 8 is 40. Hence, if the last time trains came together was 15 minutes ago, the next time will be in 25 minutes.

Question 15: C
Tiles can be added at either end of the 3 lines of 2 tiles horizontally or at either end of the 2 lines of 2 tiles vertically. This is a total of 10, but in two cases these positions are the same (at the bottom of the left hand vertical line and the top of the right hand vertical line). So the answer is $10 - 2 = 8$.

Question 16: C
Georgia is shorter than her Mum and Dad, and each of her siblings is at least as tall as Mum (and we know Mum is shorter than Dad because Ellie is between the two), so we know Georgia is the shortest. We know that Ellie, Tom and Dad are all taller than Mum, so Mum is second shortest. Ellie is shorter than Dad and Tom is taller than Dad, so we can work out that Ellie must be third shortest.

Question 17: A
Danielle must be sat next to Caitlin. Bella must be sat next to the teaching assistant. Hence these two pairs must sit in different rows. One pair must be sat at the front with Ashley, and the other must be sat at the back with Emily. Since the teaching assistant has to sit on the left, this must mean that Bella is sat in the middle seat and either Ashley or Emily (depending on which row they are in) is sat in the right hand seat. However, Bella cannot sit next to Emily, so this means Bella and the teaching assistant must be in the front row. So Ashley must be sat in the front right seat.

Question 18: D
We can see from the fact that all the possible answers end "AME" that the letters "AME" must be translated to the last 3 letters of the coded word, "JVN", under the code. J is the 10th letter of the alphabet so it is 9 letters on from A (V is the 21st letter of the alphabet and M is the 13th, and N is the 14th letter of the alphabet and E is the 5th, therefore these pairs are also 9 letters apart). Therefore P is the code for the letter 9 letters before it in the alphabet. P is the 16th letter of the alphabet, therefore it is the code for the 7th letter of the alphabet, G. Therefore from these solutions the only possibility for the original word is GAME.

Quantitative Reasoning Answers

Q	A	Q	A	Q	A	Q	A	Q	A
1	C	6	B	11	D	16	D	21	B
2	D	7	A	12	B	17	B		
3	D	8	A	13	A	18	D		
4	D	9	B	14	B	19	B		
5	C	10	A	15	B	20	B		

SET 1

Question 1: C

We can work out that the tax rates must fit in the following equation:

($50 x Food tax rate) + ($30 x Clothes tax rate) + $80 = $88. Only the tax rates in Casova fit correctly in this equation.

Question 2: D

To answer this question, we calculate how much the supplier will make for selling the items, by considering the tax rate in each state, and deducting it from the price accordingly.

Thus, in Bolovia, each year the supplier makes 250 x ($40/1.20 + $40/1.15 + $40/1.10 + $40/1.15) = $34,812 a year. In Asteria, each year the supplier makes 250 x ($40/1.10 + $40/1.15 + $40/1.10 + $40/1.15) = $35,572. (Note that in the case of an item being applicable to 2 tax rates, the higher rate will be charged. Thus, in Bolovia, imported clothes will be charged at the clothes tax rate of 15%, since this is higher than the imports rate.)

Thus, by moving to Asteria, the supplier will make $760 more each year. Therefore it will take 26.3 years to recover the purchase cost of $20,000.

Question 3: D

If John spends $88, he will spend £12 on tax. Thus, the tax rate is 12/88 = 13.6%.

If John shops in Asteria, the maximum tax rate he would have to pay is 10%; at Casova it would be 10%. If he spends at least $50 on food in Derivia, he pays no tax on it. Thus, he can spend a maximum of $38 on imported goods (at a maximum tax rate of 15%). This equates to a tax of $5.70 (not $12). Finally, if John spends $10 on imported goods in Bolovia – he would pay $0.50 in tax. Thus, he can spend up to $78 on clothes taxed at 15%. The tax on the clothes is therefore $11.70, giving $12.20 tax in total as the maximum. Since he pays $12 dollars tax, he shops in Bolovia.

Question 4: D

The sum of the basic prices is 100+30+10+100 = $240. Now the highest tax rate on the board is 20% (for imports to Asteria), thus the maximum tax is $240 x 1.20 = $288. However, this is impossible to attain (since if we bought everything in Asteria, the ham would be cheaper, as it is not imported and would only be taxed at the food rate). Therefore no option allows the overall price to be as high as $288, so this is the answer. Answer A) is possible if all products were bought in the state they are produced in. Answer C) is the correct answer if all products were bought in Asteria (and accounting for the reduced tax rate for the ham, which is not an import). Answer B) is possible if the ham was bought in Asteria, the caviar and orange juice were bought in Casova and the dress was bought in Bolovia.

SET 2

Question 5: C

Firstly, find the pressure it can withstand in Pascals: 200 pounds per square inch x 7,000 Pascals per pound per square inch = 1.4 million Pascals.

Then divide this by 1,000 Pa to get the depth the probe can withstand (we can see from the question that the pressure increases by 1,000 Pa for every metre depth increase):

1,400,000/1,000 = 1,400 metres into the ocean, which is 1.4 km.

Question 6: B

Calculate that the probe can drop 300,000 Pa/1,000 Pa per metre = 300 metres into the ocean before breaking.

Now rearrange the equation in the question, to make t the subject, as follows:

$2d = \sqrt{(t^3)}$

$(2d)^2 = t^3$

$t = \sqrt[3]{(2d)^2}$

Then substitute the depth into this supplied equation:

$t = \sqrt[3]{(2d)^2} = \sqrt[3]{(2 \times 300)^2} = 71$ seconds.

SET 3

Question 7: A

Calculate the amount of drug taken for each disease:

Black Trump Virus = 4 mg x 80 kg x 3 times a day x 28 days = 26.88 g

Swamp Fever = 3x80x1x7 = 1.68 g

Yellow Tick = 1x80x2x84 = 13.44 g

Red Rage = 5x80x2x21 = 16.80 g

At a quick glance, the swamp fever dosage is much lower than all the others – you can discount this and use that to save a little time if you need to.

Question 8: A

First calculate that Carol took 20.16 grams of the drug during the two courses for Yellow Tick, using the same method as for John, but using Carol's weight of 60kg. Therefore 20.16 grams (the amount left over) corresponds to the dosage for the unknown disease:

4x60x3x28 = 20.16 g, therefore the unknown disease was Black Trump Virus.

Question 9: B

The first time he takes 3 x 80 x 1 x 7 = 1.68 grams, and the second time he takes 4 x 110 x 3 x 28 = 36.96 grams. Thus the ratio is 1.68 : 36.96 = 1:22.

Question 10: A

By calculating the dose required in each of the cases, we see that the only one that is above 15.5 grams over 4 weeks is the dosage for Red Rage:

5 x 75 x 2 x 21 = 15.75 g – therefore Danny must be suffering from Red Rage.

Question 11:D

Heavier people need a higher dose. To find the maximum weight, we use the equation: 5 x weight x 2 x 21 = 10 g, where "weight" represents the maximum weight requiring a dosage of less than 10 g.

So the maximum weight to not need a dosage exceeding 10 g is = 10,000 mg/(21 days x 2 daily doses x 5 mg/kg) = 47.62 kg.

SET 4

Question 12: B

To solve this, divide the flour content by the overall mass. A quick inspection might show you that this is likely to be Madeira, which is confirmed by the calculation (250/825 = 0.3). Thus, 30% of the Madeira's total weight is flour, which is a higher percentage than for any other cake

Question 13: A

In this question, there must be one cake where: (2,600/mass of cake) = (625/mass of flour in cake). Thus, there is a number that both the mass of the cake and the mass of flour can be multiplied by, in order to get these numbers respectively.

We can see that if we multiply the mass of the sponge cake by 5, we get 2,600 g. Equally, if we multiply the mass of flour in the sponge cake recipe (125g) by 5, then we get 625 g. Thus, Sponge cake is the answer. No other cake recipe can be multiplied by a given number to get an overall weight of 2,600 g and 625 g of flour.

Question 14: B

We use 1.50+1.25+1.10+1 times the ingredients for one cake, so the wedding cake will use 4.85 times as much of the ingredients listed for one cakes. We can use this to find which of the possible answers can be the amount of sugar in the cake, i.e. the sugar called for in one recipe multiplied by 4.85.

The quickest way to do this is to divide each possible answer by 4.85, and see if the result matches the weight of sugar in any of the cakes. We see that 970 g/4.85 = 200 g, which is the amount of sugar in the chocolate cake. None of the other amounts are possible. Thus, B is the answer.

Question 15: B

A kilogram of flour costs 55 x 2/3 pence and we are using 0.25 kg, so 9.167 p worth of flour goes into a Madeira cake. For sugar, we have 0.175 kg x 70 p per kg = 12.25 p worth of sugar going into the cake.

The ratio is thus 9.167:12.25 = approx 0.75:1 = 3:4

Question 16: D

As before, the flour costs: 55 p per 1.5 kg x 2/3 x 0.2 kg = 7.3 pence.
The milk costs: 44 p per kg x 150 g/1000 g per kg = 6.6 pence.
Thus the ratio is 7.3:6.6 = 1:0.9 = 10:9.

SET 5

Question 17: B

In total, 108 people out of the 200 tested have the disease; this is 108/200 = 0.54. Thus, the answer is 54%.

Question 18: D

As the infection rate is different for men and women, the infection rates must be calculated separately and combined:

(231,768 x 0.53 women x 0.63) + (231,768 x 0.47 men x 0.45) = 126,406 to the nearest whole person.

Question 19: B

There are 45 men and 63 women in the test group who have the Kryptos virus. Thus 15 of the men and 45 of the women have visited Atlantis. As we now know that 60 people have visited Atlantis, we can see that 108 – 60 = 48 have not visited. Now we simply calculate 48 as a percentage of 108. 48/108 = 0.44. Thus, 44% of people testing positive for the Virus in Test A have *not* visited Atlantis.

Question 20: B

We can see that 20/45 men testing positive in Test A have also tested positive in Test B, so we assume that the rest were false positives stemming from the inaccuracies of Test A. We are told to assume the same proportion of false positives in the women tested, so we simply apply this fraction to the number of women testing positive in Test A. Thus, we simply calculate 63 x (20/45) = 28. Thus, we expect that 28 women actually have Kryptos Virus.

Question 21: B

In total 108 people tested positive under test A, and 49 of these tested positive under test B (using the data given in the last question). Therefore the percentage of people positive in Test A also testing positive in Test B is 49 out of 108, which is 45.4%.

Abstract Reasoning Answers

Q	A	Q	A	Q	A	Q	A	Q	A
1	B	11	A	21	A	31	B	41	C
2	Neither	12	B	22	A	32	Neither	42	A
3	A	13	Neither	23	B	33	Neither	43	A
4	A	14	A	24	B	34	A	44	D
5	Neither	15	A	25	Neither	35	A	45	D
6	B	16	B	26	B	36	B		
7	A	17	B	27	A	37	A		
8	A	18	A	28	A	38	Neither		
9	Neither	19	A	29	B	39	A		
10	Neither	20	A	30	B	40	Neither		

	Set A Rule	**Set B Rule**
Set 1	**3** Shapes are white **2** Shapes are black	**2** Shapes are white **3** Shapes are black
Set 2	**2** Shapes are white **1** Shape is black	**1** Shape is white **2** Shapes are black
Set 3	There is always a **triangle** in the **top left corner**	There is always a **quadrilateral** in the **bottom right corner**
Set 4	There are an **even** number of rectangles	There are an **odd** number of rectangles
Set 5	Each circle has at least one tangential line.	At least one circle is **intersected** by a line.
Set 6	**None** of the shapes intersect with each other.	At least **one** of the shapes intersects with another shape/line.
Set 7	The total number of dots is **10**	The total number of dots is **9**
Set 8	The four-sided shapes are in **different** section of the box.	The four-sided shapes are in the **same** section of the box.

Question 41: C
As the sequences progresses, the number of objects within each frame increases by two.

Question 42: A
The total number of objects in each panel totals seven. In the bottom half, the number of white circles increases by one each panel. In the top half, the number of black circles alternates between one and three between each panel with the grey circles making up the remainder to total seven.

Question 43: A
The sequence alternates between a 3-edged and 4-edged zig-zag. Within each zig-zag will be a black circle and a grey shape.

Question 44: D
The circle and triangle begin at the bottom-left corner and rotate clockwise throughout the panels and cycle between the colours white-grey-black. Each panel also contains two white stars and a square.

Question 45: D
The top-right and bottom-left shapes are both the same colour and alternate between white and black. The top-left and bottom-right shapes are also both the same colour and alternate between black and white.

Situational Judgement Answers

Q	A	Q	A	Q	A	Q	A
1	D	11	D	21	D	31	C
2	B	12	A	22	B	32	B
3	D	13	D	23	A	33	B
4	A	14	C	24	D	34	C
5	B	15	B	25	A	35	A
6	A	16	A	26	D	36	A
7	B	17	D	27	A	37	D
8	D	18	D	28	A	38	C
9	C	19	C	29	D	39	A
10	D	20	C	30	A	40	A

How appropriate is…? Scenarios: 1-6	A	A very appropriate thing to do
	B	Appropriate, but not ideal
	C	Inappropriate, but not awful
	D	A very inappropriate thing to do
How important is…? Scenarios: 7 & 8	A	Very important
	B	Important
	C	Of minor importance
	D	Not important at all

Scenario 1:

1. **Very inappropriate** because Jacob didn't know the entire story and this could be resolved by having a simple conversation instead. There is no need at his stage to get more people involved, especially as the doctor's behaviour has not directly affected the patient's safety or treatment.

2. **Appropriate but not ideal**, because the medical student would not be telling the doctor anything specific.

3. **Very inappropriate** because as a medical student you are a member of the health care team as well, so if there is something that is affecting the rest of the staff and patients, then it should not be ignored.

4. **Very appropriate** because the supervisor would be able to advise the student as to what they should do.

5. **Appropriate but not ideal**, whilst it makes Dr Herbert aware of the issue, it is quite confrontational and Dr Herbert may become defensive.

Scenario 2:

6.   **Very appropriate**, because Dr Walker could have gotten into a bad habit and may be unaware that he hasn't been washing his hands.

7.   **Appropriate but not ideal,** because Dr Walker may not pick up on the hint, although it might save George some awkwardness in having to ask Dr Walker directly.

8.   **Very inappropriate** because hospitals function as a team. If George is aware of something that could potentially cause patients harm, he must try to solve the issue.

9.   **Inappropriate but not awful**, because it is not addressing the situation and could make it a bigger problem than it actually is. In general, problems with doctors should be escalated to more senior doctors; problems with nurses should be escalated to more senior nurses.

10. **Very inappropriate** because Dr Walker would not have been informed and the fact that George would have witnessed it without trying to correct the problem could back fire onto him and get him into trouble if any harm were to arise.

Scenario 3:

11. **Very inappropriate** because the medical school museum would have to account for the missing bones. The bones are very valuable, and even the remains of bones could be useful as they would make them into slides or as cut sections.

12. **Very appropriate** because it acknowledges the respect to both the bones and the museum. The bones came from a real human, so cannot be treated as though they are any old piece of waste.

13. **Very inappropriate** because this would prompt an investigation and would waste a lot of money from the medical school. It would also mean that future classes may be banned from performing such projects, hindering their educational experience.

14. **Inappropriate but not awful,** because despite the fact that the curator would be informed of what occurred, it could get your colleague into trouble that could have been avoided.

15. **Appropriate but not ideal,** because the supervisor can give advice as to what to do, but it does not directly address the problem.

Scenario 4:

16. **Very appropriate** because his tutor can give him proper advice and will also be aware of any reasons behind a potentially disappointing exam mark.

17. **Very inappropriate** because it will increase tensions and result in a more stressful environment, which would hinder his progress even more.

18. **Very inappropriate** because he will end up feeling very isolated and lonely and anxious, which will also ruin both his friendships and his work progress.

19. **Inappropriate but not awful,** because he will lose out on friendships as well as go through the difficulty of finding another place to live. This could end up as a lonely option with no support network.

20. **Appropriate but not ideal** because he may find that everyone is having similar problems. It doesn't directly address the problems but Henry might find it helpful to discuss the situation with someone else.

Scenario 5:

21. **Very inappropriate** because Mark was only asked because he was mistaken for a doctor. Therefore exchanging one student for another would be an inappropriate action. Also, the other student wasn't asked.

22. **Appropriate but not ideal,** because Mark could document the results as a student and write exactly who he was. Students are allowed to perform tests, just not to administer medication.

23. **Very appropriate** because it will alert the doctor as to his mistake, and the student can be advised appropriately.

24. **Very inappropriate** because the test needs to be done, and the doctor would assume that the test had been done. Therefore the patient could be left waiting for a long time.

25. **Very appropriate** because they might be better qualified to do the test.

Scenario 6:

26. **Very inappropriate** because it is dishonest and if the truth were to emerge, she could be expelled from the dental school for such an act.
27. **Very appropriate** because they can support her and help her organise the rest of her revision.
28. **Very appropriate** because it can reassure her and give her more confidence as most people would probably feel similar.
29. **Very inappropriate** because this is also dishonest and therefore, unprofessional.
30. **Very appropriate** because she might just pass her exams and surprise herself.

Scenario 7:

31. **Of minor importance** because the task should be joint effort.
32. **Important,** because this means that their grade is significant, and Sean will want to do as well as possible.
33. **Important** because they should learn how to work together to prevent future problems with their group work.
34. **Of minor importance**- although this might be troubling Daniel, he shouldn't let his social life affect his work life.
35. **Very important** because Daniel maybe used to not pulling his weight, and will have to be informed that he needs to contribute more.

Scenario 8:

36. **Very important**, because if the essay counts towards Tanya's final mark for the year then she would want to do very well.
37. **Not important at all** – different people learn in different ways and every clinic is different.
38. Of minor importance, because Dr Garg will be assessing her at the end of the term.
39. **Very important**, because if this opportunity is available at another time, then missing this particular clinic is not particularly disastrous for Tanya's learning.
40. **Very important,** because Tanya can complete the essay in time for the clinic, then she would not be compromising her learning.

THE BMAT

The Basics

What is the BMAT?

The BioMedical Admissions Test (BMAT) is a 2-hour written exam for dental, medicine and veterinary students who are applying for certain universities. In 2016 Leeds University became the first dental school in the UK to have the BMAT a requisition for dentistry as part of their application process, so be wary that other universities may follow suit in the future. Currently only Leeds dental school require this test for their dental application process (At the time of print).

What does the BMAT consist of?

Section	SKILLS TESTED	Questions	Timing
ONE	Problem-solving skills, including numerical and spatial reasoning. Critical thinking skills, including understanding argument and reasoning using everyday language.	35 MCQs	60 minutes
TWO	Ability to recall, understand and apply GCSE level principles of biology, chemistry, physics and maths. Usually the section that students find the hardest.	27 MCQs	30 minutes
THREE	Ability to organise ideas in a clear and concise manner, and communicate them effectively in writing. Questions are usually but not necessarily dental.	One essay from four	30 minutes

Why is the BMAT used?

Dental, medical and veterinary applicants tend to be a bright bunch and therefore usually have excellent grades. This means that competition is fierce – meaning that the universities must use the BMAT to help differentiate between applicants.

When do I sit BMAT?

The BMAT takes place in September or November each year. For 2018 the dates are as follows;
25th June – Registration opens
12th August – Registration closing date
1st September – BMAT September test date
21st September – BMAT results released
20th October – Final date to share results with selected universities

Can I re-sit the BMAT?

No, you can only sit the BMAT once per admissions cycle.

Where do I sit the BMAT?

You can usually sit the BMAT at your school or college (ask your exams officer for more information). Alternatively, if your school isn't a registered test centre or you're not attending a school or college, you can sit the BMAT at an authorised test centre.

Who has to sit the BMAT?

For 2018 entry, applicants to the following universities must sit the BMAT:

University	Course
University of Cambridge	Medicine and Veterinary Medicine
University of Oxford	Medicine, Graduate Medicine, Biomedical Science
University College London	Medicine
Imperial College London	Medicine, Graduate Medicine, Biomedical Science
Brighton and Sussex	Medicine
University of Leeds	Medicine, Dentistry
Lancaster University	Medicine
Keele University	Medicine
Royal Veterinary College	Veterinary Medicine
Lee Kong Chian (Singapore)	Medicine
Melbourne University (Aus)	Medical of Doctors Surgery
Mahidol University (Thailand)	Medicine
Thammasat University (Thai)	Medicine

Do I have to re-sit the BMAT if I reapply?

You only need to re-sit the BMAT if you are applying to a university that requires it. You cannot use your score from any previous attempts.

How is the BMAT Scored?

Section 1 and Section 2 are marked on a scale of 0 to 9. Generally, 5 is an average score, 6 is good, and 7 is excellent. Very few people (less than 5%) get more than 8.

The marks for sections 1 + 2 show a normal distribution with a large range. The important thing to note is that the difference between a score of 5.0 and 6.5+ is often only 3-4 questions. Thus, you can see that even small improvements in your raw score will lead to massive improvements when they are scaled.

Section 3 is marked on 2 scales:
– A-E for Quality of English
– 0-5 for Strength of Argument

The marks for sections 3 show a normal distribution for the strength of argument; the average mark for the strength of argument is between 3 – 3.5.

The quality of English marks are negatively skewed distribution. I.e. the vast majority of students will score A or B for quality of English. The ones that don't tend to be students who are not fluent in English.

This effectively means that the letter score is used to flag students who have a comparatively weaker grasp of English- i.e. it is a test of competence rather than excellence like the rest of the BMAT. This effectively means that if you get a C or below, admissions tutors are more likely to scrutinise your essay than otherwise.

Finally, section 3 is marked by two different examiners. If there is a large discrepancy between their marks, it is marked by a third examiner.

When do I get my results?

The BMAT results are usually released to universities in mid-late November and then to students in late November. Normally, you'll be sent them via email or you'll get a login that will allow you to access them online.

How is the BMAT used?

Cambridge: Cambridge interviews more than 90% of students who apply so the BMAT score isn't vital for making the interview shortlist. However, it can play a huge role in the final decision – for example, 50% of overall marks for your application may be allocated to the BMAT. Thus, it's essential you find out as much information about the college you're applying to.

Oxford: Oxford typically receives thousands of applications each year and they use the BMAT to shortlist students for interview. Typically, 450 students are invited for interview for 150 places. Thus, if you get offered an interview- you are doing very well! Oxford centralise their short listing process and use an algorithm that uses your % A*s at GCSE along with your BMAT score to rank all their applicants of which the top are invited to interview. BMAT sections 1 + 2 count for 40% each of your BMAT score whilst section 3 counts for 20% [the strength of argument (number) contributes to 13.3% and the quality of English (letter) makes up the remaining 6.7%].

UCL: UCL make offers based on all components of the application and whilst the BMAT is important there is no magic threshold that you need to meet in order to guarantee an interview. Applicants with higher BMAT scores tend to be interviewed earlier in the year.

Imperial: Imperial employs a BMAT threshold to shortlist for interview. This exact threshold changes every year but in the past has been approximately 4.5-5.0 for sections 1 + 2 and 2.5 B for section 3.

Leeds: The BMAT contributes to 15% of your academic score at Leeds. You will be allocated marks based on your rank in the BMAT. Thus, applicants in the top 20% of the BMAT will get the full quota of marks for their application and the bottom 20% will get the lowest possible mark for their application. Thus, you can still get an interview if you perform poorly in the BMAT (it' just much harder!). Leeds will calculate your BMAT score by attributing 40% to section 1, 40% to section 2 and 20% to section 3 (lower weighting as it can come up during the interview).

Brighton: Brighton started using the BMAT in 2014 so little is known about how they use it in their decision making process. They state on their website that it "may also be used as a final discriminator if needed after interview."

Royal Veterinary College: It is unclear how the RVC use the BMAT- it has influenced applications both before and after interview and it's likely that they use it on a case-by-case basis rather than as an arbitrary cut-off.

General Advice

Start Early

It is much easier to prepare if you practice little and often. Start your preparation well in advance; ideally by mid September but at the latest by early October. This way you will have plenty of time to complete as many papers as you wish to feel comfortable and won't have to panic and cram just before the test, which is a much less effective and more stressful way to learn. In general, an early start will give you the opportunity to identify the complex issues and work at your own pace.

Prioritise

Some questions in sections 1 + 2 can be long and complex – and given the intense time pressure you need to know your limits. It is essential that you don't get stuck with very difficult questions. If a question looks particularly long or complex, mark it for review and move on. You don't want to be caught 5 questions short at the end just because you took more than 3 minutes in answering a challenging multi-step physics question. If a question is taking too long, choose a sensible answer and move on. Remember that each question carries equal weighting and therefore, you should adjust your timing in accordingly. With practice and discipline, you can get very good at this and learn to maximise your efficiency.

Positive Marking

There are no penalties for incorrect answers in the BMAT; you will gain one for each right answer and will not get one for each wrong or unanswered one. This provides you with the luxury that you can always guess should you absolutely be not able to figure out the right answer for a question or run behind time. Since each question provides you with 4 to 6 possible answers, you have a 16-25% chance of guessing correctly. Therefore, if you aren't sure (and are running short of time), then make an educated guess and move on. Before 'guessing' you should try to eliminate a couple of answers to increase your chances of getting the question correct. For example, if a question has 5 options and you manage to eliminate 2 options- your chances of getting the question increase from 20% to 33%!

Avoid losing easy marks on other questions because of poor exam technique. Similarly, if you have failed to finish the exam, take the last 10 seconds to guess the remaining questions to at least give yourself a chance of getting them right.

Practice

This is the best way of familiarising yourself with the style of questions and the timing for this section. Although the BMAT tests only GCSE level knowledge, you are unlikely to be familiar with the style of questions in all 3 sections when you first encounter them. Therefore, you want to be comfortable at using this before you sit the test.

Practising questions will put you at ease and make you more comfortable with the exam. The more comfortable you are, the less you will panic on the test day and the more likely you are to score highly. Initially, work through the questions at your own pace, and spend time carefully reading the questions and looking at any additional data. When it becomes closer to the test, **make sure you practice the questions under exam conditions**.

Past Papers

Official past papers and answers from 2003 onwards are freely available online on our website at **www.uniadmissions.co.uk/bmat-past-papers** and once you've worked your way through the questions in this book, you are highly advised to attempt as many of them as you can (ideally at least 5). If you get stuck, you can also get access to fully worked solutions to all the past papers. Keep in mind that the specification was changed in 2009 so some things asked in earlier papers may not be representative of the content that is currently examinable in the BMAT. In general, **it is worth doing at least all the papers from 2009 onwards**. Time permitting; you can work backwards from 2009 although there is little point doing the section 3 essays pre-2009 as they are significantly different to the current style of essays.

Repeat Questions

When checking through answers, pay particular attention to questions you have got wrong. If there is a worked answer, look through that carefully until you feel confident that you understand the reasoning, and then repeat the question without help to check that you can do it. If only the answer is given, have another look at the question and try to work out why that answer is correct. This is the best way to learn from your mistakes, and means you are less likely to make similar mistakes when it comes to the test. The same applies for questions which you were unsure of and made an educated guess which was correct, even if you got it right. When working through this book, **make sure you highlight any questions you are unsure of**, this means you know to spend more time looking over them once marked.

No Calculators

You aren't permitted to use calculators in the BMAT – thus, it is essential that you have strong numerical skills. For instance, you should be able to rapidly convert between percentages, decimals and fractions. You will seldom get questions that would require calculators but you would be expected to be able to arrive at a sensible estimate. Consider for example:

Estimate 3.962 x 2.322;

3.962 is approximately 4 and 2.323 is approximately 2.33 = 7/3.

Thus, $3.962 \times 2.322 \approx 4 \times \frac{7}{3} = \frac{28}{3} = 9.33$

Since you will rarely be asked to perform difficult calculations, you can use this as a signpost of if you are tackling a question correctly. For example, when solving a physics question, you end up having to divide 8,079 by 357- this should raise alarm bells as calculations in the BMAT are rarely this difficult.

Top tip! In general, students tend to improve the fastest in section 2 and slowest in section 1; section 3 usually falls somewhere in the middle. Thus, if you have very little time left, it's best to prioritise section 2.

A word on timing...

"If you had all day to do your BMAT, you would get 100%. But you don't."

Whilst this isn't completely true, it illustrates a very important point. Once you've practiced and know how to answer the questions, the clock is your biggest enemy. This seemingly obvious statement has one very important consequence. **The way to improve your BMAT score is to improve your speed.** There is no magic bullet. But there are a great number of techniques that, with practice, will give you significant time gains, allowing you to answer more questions and score more marks.

Timing is tight throughout the BMAT – **mastering timing is the first key to success**. Some candidates choose to work as quickly as possible to save up time at the end to check back, but this is generally not the best way to do it. BMAT questions can have a lot of information in them – each time you start answering a question it takes time to get familiar with the instructions and information. By splitting the question into two sessions (the first run-through and the return-to-check) you double the amount of time you spend on familiarising yourself with the data, as you have to do it twice instead of only once. This costs valuable time. In addition, candidates who do check back may spend 2–3 minutes doing so and yet not make any actual changes. Whilst this can be reassuring, it is a false reassurance as it is unlikely to have a significant effect on your actual score. Therefore it is usually best to pace yourself very steadily, aiming to spend the same amount of time on each question and finish the final question in a section just as time runs out. This reduces the time spent on re-familiarising with questions and maximises the time spent on the first attempt, gaining more marks.

It is essential that you don't get stuck with the hardest questions – no doubt there will be some. In the time spent answering only one of these you may miss out on answering three easier questions. If a question is taking too long, choose a sensible answer and move on. Never see this as giving up or in any way failing, rather it is the smart way to approach a test with a tight time limit. With practice and discipline, you can get very good at this and learn to maximise your efficiency. It is not about being a hero and aiming for full marks – this is almost impossible and very much unnecessary (even Oxbridge will regard any score higher than 7 as exceptional). It is about maximising your efficiency and gaining the maximum possible number of marks within the time you have.

Top tip! Ensure that you take a watch that can show you the time in seconds into the exam. This will allow you have a much more accurate idea of the time you're spending on a question. In general, if you've spent >150 seconds on a section 1 question or >90 seconds on a section 2 questions – move on regardless of how close you think you are to solving it.

Use the Options:

Some questions may try to overload you with information. When presented with large tables and data, it's essential you look at the answer options so you can focus your mind. This can allow you to reach the correct answer a lot more quickly. Consider the example below:

The table below shows the results of a study investigating antibiotic resistance in staphylococcus populations. A single staphylococcus bacterium is chosen at random from a similar population. Resistance to any one antibiotic is independent of resistance to others.

Calculate the probability that the bacterium selected will be resistant to all four drugs.

A   1 in 10^6
B   1 in 10^{12}
C   1 in 10^{20}
D   1 in 10^{25}
E   1 in 10^{30}
F   1 in 10^{35}

Antibiotic	Number of Bacteria tested	Number of Resistant Bacteria
Benzyl-penicillin	10^{11}	98
Chloramphenicol	10^9	1200
Metronidazole	10^8	256
Erythromycin	10^5	2

Looking at the options first makes it obvious that there is **no need to calculate exact values**- only in powers of 10. This makes your life a lot easier. If you hadn't noticed this, you might have spent well over 90 seconds trying to calculate the exact value when it wasn't even being asked for.

In other cases, you may actually be able to use the options to arrive at the solution quicker than if you had tried to solve the question as you normally would. Consider the example below:

A region is defined by the two inequalities: $x - y^2 > 1$ and $xy > 1$. Which of the following points is in the defined region?

A.   (10,3)
B.   (10,2)
C.   (-10,3)
D.   (-10,2)
E.   (-10,-3)

Whilst it's possible to solve this question both algebraically or graphically by manipulating the identities, by far **the quickest way is to actually use the options**. Note that options C, D and E violate the second inequality, narrowing down to answer to either A or B. For A: $10 - 3^2 = 1$ and thus this point is on the boundary of the defined region and not actually in the region. Thus the answer is B (as 10-4 = 6 > 1.)

In general, it pays dividends to look at the options briefly and see if they can be help you arrive at the question more quickly. Get into this habit early – it may feel unnatural at first but it's guaranteed to save you time in the long run.

Keywords

If you're stuck on a question; pay particular attention to the options that contain key modifiers like "**always**", "**only**", "**all**" as examiners like using them to test if there are any gaps in your knowledge. E.g. the statement "arteries carry oxygenated blood" would normally be true; "All arteries carry oxygenated blood" would be false because the pulmonary artery carries deoxygenated blood.

SECTION 1

This is the first section of the BMAT and as you walk in, it is inevitable that you will feel nervous. Make sure that you have been to the toilet because once it starts you cannot simply pause and go. Take a few deep breaths and calm yourself down. Remember that panicking will not help and may negatively affect your marks- so try and avoid this as much as possible.

You have one hour to answer 35 questions in section 1. The questions fall into three categories:
-   Problem solving
-   Data handling
-   Critical thinking

Whilst this section of the BMAT is renowned for being difficult to prepare for, there are powerful shortcuts and techniques that you can use to save valuable time on these types of questions.

You have approximately 100 seconds per question; this may sound like a lot but given that you're often required to read and analyse passages or graphs- it can often not be enough. Nevertheless, this section is not as time pressured as section 2 so most students usually finish the majority of questions in time. However, some questions in this section are very tricky and can be a big drain on your limited time. **The people who fail to complete section 1 are those who get bogged down on a particular question**.

Therefore, it is vital that you start to get a feel for which questions are going to be easy and quick to do and which ones should be left till the end. The best way to do this is through practice and the questions in this book will offer extensive opportunities for you to do so.

SECTION 1: Critical Thinking Questions

BMAT Critical thinking questions require you to understand the constituents of a good argument and be able to pick them apart. The majority of BMAT Critical thinking questions tend to fall into 3 major categories:

1. Identifying Conclusions
2. Identifying Assumptions + Flaws
3. Strengthening and Weakening arguments

Having a good grasp of language and being able to filter unnecessary information quickly and efficiently is a vital skill in dental school – you simply do not have the time to sit and read vast numbers of textbooks cover to cover, you need to be able to filter the information and realise which part is important and this will contribute to your success in your studies. Similarly, when you have qualified and are on the wards, you need to be able to pick out key information from patient notes and make healthcare decisions from them, so getting to grips with verbal reasoning goes a long way and do not underestimate its importance.

Only use the Passage

Your answer must only be based on the information available in the passage. Do not try and guess the answer based on your general knowledge as this can be a trap. For example, if the passage says that spring is followed by winter, then take this as true even though you know that spring is followed by summer.

> *Top tip!* Though it might initially sound counter-intuitive, it is often best to read the question *before* reading the passage. Then you'll have a much better idea of what you're looking for and are therefore more likely to find it quicker.

Take your time

Unlike the problem solving questions, critical thinking questions are less time pressured. Most of the passages are well below 300 words and therefore don't take long to read and process (unlike the UKCAT in which you should skim read passages). Thus, your aim should be to understand the intricacies of the passage and identify key information so that you don't miss key information and lose easy marks.

Identifying Conclusions

Students struggle with these type of questions because they confuse a premise for a conclusion. For clarities sake:

- A **Conclusion** is a summary of the arguments being made and is usually explicitly stated or heavily implied.
- A **Premise** is a statement from which another statement can be inferred or follows as a conclusion.

Hence a conclusion is shown/implied/proven by a premise. Similarly, a premise shows/indicates/establishes a conclusion. Consider for example: *My mom, being a woman, is clever as all women are clever.*

Premise 1: My mom is a woman + **Premise 2:** Women are clever = **Conclusion:** My mom is clever.

This is fairly straightforward as it's a very short passage and the conclusion is explicitly stated. Sometimes the latter may not happen. Consider: *My mom is a woman and all women are clever.*
Here, whilst the conclusion is not explicitly being stated, both premises still stand and can be used to reach the same conclusion.

You may sometimes be asked to identify if any of the options cannot be "reliably concluded". This is effectively asking you to identify why an option **cannot** be the conclusion. There are many reasons why but the most common ones are:

1. Over-generalising: *My mom is clever therefore all women are clever.*
2. Being too specific: All kids like candy thus my son also likes candy.
3. Confusing Correlation vs. Causation: *Lung cancer is much more likely in patients who drink water. Hence, water causes lung cancer.*
4. Confusing Cause and Effect: *Lung cancer patients tend to smoke so it follows that having lung cancer must make people want to smoke.*

Note how conjunctives like hence, thus, therefore and it follows give you a clue as to when a conclusion is being stated. More examples of these include: "it follows that, implies that, whence, entails that".
Similarly, words like "because, as indicated by, in that, given that, due to the fact that" usually identify premises.

Assumptions + Flaws:

Other types of critical thinking questions may require you to identify assumptions and flaws in a passage's reasoning. Before proceeding it is useful to define both:

- An assumption is a reasonable assertion that can be made on the basis of the available evidence.
- A flaw is an element of an argument which is inconsistent to the rest of the available evidence. It undermines the crucial components of the overall argument being made.

Consider for example: *My mum is clever because all dentists are clever.*

Premise 1: Dentists are clever. **Assumption:** My mum is a dentist. **Conclusion:** My mum is clever.

Note that the conclusion follows naturally even though there is only one premise because of the assumption. The argument relies on the assumption to work. Thus, if you are unsure if an option you have is an assumption or not, just ask yourself:

1) *Is it in the passage?* If the answer is **no** then proceed to ask:
2) *Does the conclusion rely on this piece of information in order to work?* – If the answer is **yes** – then you've identified an assumption.

You may sometimes be asked to identify flaws in an argument – it is important to be aware of the types of flaws to look out for. In general, these are broadly similar to the ones discussed earlier in the conclusion section (over-generalising, being too specific, confusing cause and effect, confusing correlation and causation). Remember that an assumption may also be a flaw.

For example consider again: *My mum is clever because all dentists are clever.*

What if the mother was not actually a dentist? The argument would then breakdown as the assumption would be incorrect or **flawed**.

> *Top tip!* Don't get confused between premises and assumptions. A **premise** is a statement that is explicitly stated in the passage. An **assumption** is an inference that is made from the passage.

Critical Thinking Questions

Question 1-6 are based on the passage below:

People have tried to elucidate the differences between the different genders for many years. Are they societal pressures or genetic differences? In the past it has always been assumed that it was programmed into our DNA to act in a certain more masculine or feminine way but now evidence has emerged that may show it is not our genetics that determines the way we act, but that society pre-programmes us into gender identification. Whilst it is generally acknowledged that not all boys and girls are the same, why is it that most young boys like to play with trucks and diggers whilst young girls prefer dollies and pink?

The society we live in has always been an important factor in our identity, take cultural differences; the language we speak the food we eat, the clothes we wear. All of these factors influence our identity. New research finds that the people around us may prove to be the biggest influence on our gender behaviour. It shows our parents buying gendered toys may have a much bigger influence than the genes they gave us. Girls are being programmed to like the same things as their mothers and this has lasting effects on their personality. Young girls and boys are forced into their gender stereotypes through the clothes they are bought, the hairstyle they wear and the toys they play with.

The power of society to influence gender behaviour explains the cases where children have been born with different external sex organs to those that would match their sex determining chromosomes. Despite the influence of their DNA they identify to the gender they have always been told they are. Once the difference has been detected, how then are they ever to feel comfortable in their own skin? The only way to prevent society having such a large influence on gender identity is to allow children to express themselves, wear what they want and play with what they want without fear of not fitting in.

Question 1:
What is the main conclusion from the first paragraph?
A.   Society controls gender behaviour.
B.   People are different based on their gender.
C.   DNA programmes how we act.
D.   Boys do not like the same things as girls because of their genes.

Question 2:
Which of the following, if true, points out the flaw in the first paragraph's argument?
A.   Not all boys like trucks.
B.   Genes control the production of hormones.
C.   Differences in gender may be due to an equal combination of society and genes.
D.   Some girls like trucks.

Question 3:
According to the statement, how can culture affect identity?
A.   Culture can influence what we wear and how we speak.
B.   Our parents act the way they do because of culture.
C.   Culture affects our genetics.
D.   Culture usually relates to where we live.

Question 4:

Which of these is most implied by the statement?

A. Children usually identify with the gender they appear to be.
B. Children are programmed to like the things they do by their DNA.
C. Girls like dollies and pink because their mothers do.
D. It is wrong for boys to have long hair like girls.

Question 5:

What does the statement say is the best way to prevent gender stereotyping?

A. Mothers spending more time with their sons.
B. Parents buying gender-neutral clothes for their children.
C. Allowing children to act how they want.
D. Not telling children if they have different sex organs.

Question 6:

What, according to the statement is the biggest problem for children born with different external sex organs to those which match their sex chromosomes?

A. They may have other problems with their DNA.
B. Society may not accept them for who they are.
C. They may wish to be another gender.
D. They are not the gender they are treated as which can be distressing.

Questions 7-11 are based on the passage below:

New evidence has emerged that the most important factor in a child's development could be their napping routine. It has come to light that regular napping could well be the deciding factor for determining toddlers' memory and learning abilities. The new countrywide survey of 1000 toddlers, all born in the same year showed around 75% had regular 30-minute naps. Parents cited the benefits of their child having a regular routine (including meal times) such as decreased irritability, and stated the only downfall of occasional problems with sleeping at night. Research indicating that toddlers were 10% more likely to suffer regular night-time sleeping disturbances when they regularly napped supported the parent's view.

Those who regularly took 30-minute naps were more than twice as likely to remember simple words such as those of new toys than their non-napping counterparts, who also had higher incidences of memory impairment, behavioural problems and learning difficulties. Toddlers who regularly had 30 minute naps were tested on whether they were able recall the names of new objects the following day, compared to a control group who did not regularly nap. These potential links between napping and memory, behaviour and learning ability provides exciting new evidence in the field of child development.

Question 7:

If in 100 toddlers 5% who did not nap were able to remember a new teddy's name, how many who had napped would be expected to remember?

A. 8   B. 9   C. 10   D. 12

Question 8:

Assuming that the incidence of night-time sleeping disturbances is the same in for all toddlers independent of all characteristics other than napping, what is the percentage of toddlers who suffer regular night-time sleeping disturbances as a result of napping?

A. 7.5%   B. 10%   C. 14%   D. 20%   E. 50%

Question 9:

Using the information from the passage above, which of the following is the most plausible alternative reason for the link between memory and napping?

A. Children who have bad memory abilities are also likely to have trouble sleeping.

B. Children who regularly nap, are born with better memories.

C. Children who do not nap were unable to concentrate on the memory testing exercises for the study.

D. Parents who enforce a napping routine are more likely to conduct memory exercises with their children.

Question 10:

Which of the following is most strongly indicated?

A. Families have more enjoyable meal times when their toddlers regularly nap.

B. Toddlers have better routines when they nap.

C. Parents enforce napping to improve their toddlers' memory ability.

D. Napping is important for parents' routines.

Question 11:

Which of the following, if true, would strengthen the conclusion that there is a causal link between regular napping and improved memory in toddlers?

A. Improved memory is also associated with regular mealtimes.

B. Parents who enforce regular napping are more inclined to include their children in studies.

C. Toddlers' memory development is so rapid that even a few weeks can make a difference to performance.

D. Among toddler playgroups where napping incidence is higher and more consistent memory performance is significantly improved compared to those that do not.

Question 12:

Tom's father says to him: 'You must work for your A-levels. That is the best way to do well in your A-level exams. If you work especially hard for Geography, you will definitely succeed in your Geography A-level exam'.

Which of the following is the best statement Tom could say to prove a flaw in his father's argument?

A. 'It takes me longer to study for my History exam, so I should prioritise that.'

B. 'I do not have to work hard to do well in my Geography A-level.'

C. 'Just because I work hard, does not mean I will do well in my A-levels.'

D. 'You are putting too much importance on studying for A-levels.'

Question 13:

Today the NHS is increasingly struggling to be financially viable. In the future, the NHS may have to reduce the services it cannot afford. The NHS is supported by government funds, which come from those who pay tax in the UK. Recently the NHS has been criticised for allowing fertility treatments to be free, as many people believe these are not important and should not be paid for when there is not enough money to pay the dentists and nurses.

Which of the following is the most accurate conclusion of the statement above?

A. Only taxpayers should decide where the NHS spends its money.

B. Dentists and nurses should be better paid.

C. The NHS should stop free fertility treatments.

D. Fertility treatments may have to be cut if finances do not improve.

Question 14:

'We should allow people to drive as fast as they want. By allowing drivers to drive at fast speeds, through natural selection the most dangerous drivers will kill only themselves in car accidents. They will not have children -only safe people will reproduce and eventually the population will only consist of safe drivers.'

Which one of the following, if true, most weakens the above argument?

A. Dangerous drivers harm others more often than themselves by driving too fast.

B. Dangerous drivers may produce children who are safe drivers.

C. The process of natural selection takes a long time.

D. Some drivers break speed limits anyway.

SECTION 1: Problem Solving Questions

Section 1 problem solving questions are arguably the hardest to prepare for. However, there are some useful techniques you can employ to solve some types of questions much more quickly:

Construct Equations

Some of the problems in Section 1 are quite complex and you'll need to be comfortable with turning prose into equations and manipulating them. For example, when you read "Mark is twice as old as Jon" – this should immediately register as M = 2J. Once you get comfortable forming equations, you can start to approach some of the harder questions in this book (and past papers) which may require you to form and solve simultaneous equations. Consider the example:

Nick has a sleigh that contains toy horses and clowns and counts 44 heads and 132 legs in his sleigh. Given that horses have one head and four legs, and clowns have one head and two legs, calculate the difference between the number of horses and clowns.

A. 0
B. 5
C. 22
D. 28
E. 132
F. More information is needed.

To start with, let C= Clowns and H= Horses.
For Heads: $C + H = 44$; For Legs: $2C + 4H = 132$
This now sets up your two equations that you can solve simultaneously.
$C = 44 - H$ so $2(44 - H) + 4H = 132$
Thus, $88 - 2H + 4H = 132$;
Therefore, $2H = 44$; $H = 22$
Substitute back in to give $C = 44 - H = 44 - 22 = 22$
Thus the difference between horses and clowns $= C - H = 22 - 22 = 0$

It's important you are able to do these types of questions quickly (and **without resorting to trial & error** as they are commonplace in section 1.

Spatial Reasoning

There are usually 1-2 spatial reasoning questions every year. They usually give nets for a shape or a patterned cuboid and ask which options are possible rotations. Unfortunately, they are extremely difficult to prepare for because the skills necessary to solve these types of questions can take a very long time to improve. The best thing you can do to prepare is to familiarise yourself with the basics of how cube nets work and what the effect of transformations are e.g. what happens if a shape is reflected in a mirror etc.

It is also a good idea to try to learn to draw basic shapes like cubes from multiple angles if you can't do so already. Finally, remember that if the shape is straightforward like a cube, it might be easier for you to draw a net, cut it out and fold it yourself to see which of the options are possible.

Diagrams

When a question asks about timetables, orders or sequences, draw out diagrams. By doing this, you can organise your thoughts and help make sense of the question.

"Mordor is West of Gondor but East of Rivendale. Lorien is midway between Gondor and Mordor. Erebus is West of Mordor. Eden is not East of Gondor."

*Which of the following **cannot** be concluded?*

A. Lorien is East of Erebus and Mordor.
B. Mordor is West of Gondor and East of Erebus.
C. Rivendale is west of Lorien and Gondor.
D. Gondor is East of Mordor and East of Lorien
E. Erebus is West of Mordor and West of Rivendale.

Whilst it is possible to solve this in your head, it becomes much more manageable if you draw a quick diagram and plot the positions of each town:

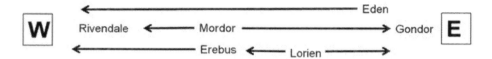

Now, it's a simple case of going through each option and seeing if it is correct according to the diagram. You can now easily see that Option E- Erebus cannot be west of Rivendale.

Don't feel that you have to restrict yourself to linear diagrams like this either – for some questions you may need to draw tables or even Venn diagrams. Consider the example:

Slifers and Osiris are not legendary. Krakens and Minotaurs are legendary. Minotaurs and Lords are both divine. Humans are neither legendary nor divine.

A. Krakens may be only legendary or legendary and divine.
B. Humans are not divine.
C. Slifers are only divine.
D. Osiris may be divine.
E. Humans and Slifers are the same in terms of both qualities.

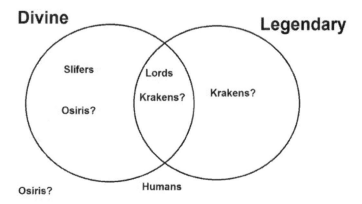

Constructing a Venn diagram allows us to quickly see that the position of Osiris and Krakens aren't certain. Thus, A and D must be true. Humans are neither so B is true. Krakens may be divine so A is true. E cannot be concluded as Slifers are divine but are humans are not. Thus, E is False.

Problem Solving Questions

Question 15:

Pilbury is south of Westside, which is south of Harrington. Twotown is north of Pilbury and Crewville but not further north than Westside. Crewville is:

F. South of Westside, Pilbury and Harrington but not necessarily Twotown.
G. North of Pilbury, and Westside.
H. South of Westside and Twotown, but north of Pilbury.
I. South of Westside, Harrington and Twotown but not necessarily Pilbury.
J. South of Harrington, Westside, Twotown and Pilbury.

Question 16:

The hospital coordinator is making the rota for the ward for next week; two of Drs Evans, James and Luca must be working on weekdays, none of them on Sundays and all of them on Saturdays. Dr Evans works 4 days a week including Mondays and Fridays. Dr Luca cannot work Monday or Thursday. Only Dr James can work 4 days consecutively, but he cannot do 5.

What days does Dr James work?
F. Saturday, Sunday and Monday.
G. Monday, Tuesday, Wednesday, Thursday and Saturday.
H. Monday, Thursday Friday and Saturday.
I. Tuesday, Wednesday, Friday and Saturday.
J. Monday, Tuesday, Wednesday, Thursday and Friday.

Question 17:

Michael, a taxi driver, charges a call out rate and a rate per mile for taxi rides. For a 4 mile ride he charges £11, and for a 5 mile ride, £13.

How much does he charge for a 9-mile ride?
A. £15   B. £17   C. £19   D. £20   E. £21

Question 18:

Goblins and trolls are not magical. Fairies and goblins are both mythical. Elves and fairies are magical. Gnomes are neither mythical nor magical.

Which of the following is **FALSE**?
A. Elves may be only magical or magical and mythical.
B. Gnomes are not mythical.
C. Goblins are only mythical.
D. Trolls may be mythical.
E. Gnomes and goblins are the same in terms of both qualities.

Question 19:

Jessica runs a small business making bespoke wall tiles. She has just had a rush order for 100 tiles placed that must be ready for today at 7pm. The client wants the tiles packed all together, a process which will take 15 minutes. Only 50 tiles can go in the kiln at any point and they must be put in the kiln to heat for 45 minutes. The tiles then sit in the kiln to cool before they can be packed, a process which takes 20 minutes. While tiles are in the kiln Jessica is able to decorate more tiles at a rate of 1 tile per minute.

What is the latest time Jessica can start making the tiles?
A. 2:55pm   B. 3:15pm   C. 3:30pm   D. 3:45pm

Question 20:

Pain nerve impulses are twice as fast as normal touch impulses. If Yun touches a boiling hot pan this message reaches her brain, 1 metre away, in 1 millisecond.

What is the speed of a normal touch impulse?

A. 5 m/s   B. 20 m/s   C. 50 m/s   D. 200m/s   E. 500 m/s

Question 21:

A woman has two children Melissa and Jack, yearly, their birthdays are 3 months apart, both being on the 22nd. The woman wishes to continue the trend of her children's names beginning with the same letter as the month they were born. If her next child, Alina is born on the 22nd 2 months after Jack's birthday, how many months after Alina is born will Melissa have her next birthday?

A. 2 months   B. 4 months   C. 5 months   D. 6 months   E. 7 months

Question 22:

Policemen work in pairs. PC Carter, PC Dirk, PC Adams and PC Bryan must work together but not for more than seven days in a row, which PC Adams and PC Bryan now have. PC Dirk has worked with PC Carter for 3 days in a row. PC Carter does not want to work with PC Adams if it can be avoided.

Who should work with PC Bryan?

A. PC Carter
B. PC Dirk
C. PC Adams
D. Nobody is available under the guidelines above.

Question 23:

My hair-dressers charges £30 for a haircut, £50 for a cut and blow-dry, and £60 for a full hair dye. They also do manicures, of which the first costs £15, and includes a bottle of nail polish, but are subsequently reduced by £5 if I bring my bottle of polish. The price is reduced by 10% if I book and pay for the next 5 appointments in advance and by 15% if I book at least the next 10.

I want to pay for my next 5 cut and blow-dry appointments, as well as for my next 3 manicures. How much will it cost?

A. £170   B. £255   C. £260   D. £285   E. £305

Question 24:

Alex, Bertha, David, Gemma, Charlie, Elena and Frankie are all members of the same family consisting of three children, two of whom, Frankie and Gemma are girls. No other assumption of gender based on name can be established. There are also four adults. Alex is a dentist and is David's brother. One of them is married to Elena, and they have two children. Bertha is married to David; Gemma is their child.

Who is Charlie?

A. Alex's daughter
B. Frankie's father
C. Gemma's brother
D. Elena's son
E. Gemma's sister

For 800 more BMAT practice questions check the *Ultimate BMAT Guide* – flick to the back to get a free copy.

SECTION 1: Data Analysis

Data analysis questions show a great variation in type and difficulty. The best way to improve with these questions is to do lots of practice questions in order to familiarise yourself with the style of questions.

Options First

Despite the fact that you may have lots of data to contend with, the rule about looking at the options first still stands in this section. This will allow you to register what type of calculation you are required to make and what data you might need to look at for this. Remember, Options → Question → Data/Passage.

Working with Numbers

Percentages frequently make an appearance in this section and it's vital that you're able to work comfortably with them. For example, you should be comfortable increasing and decreasing by percentages, and working out inverse percentages too. When dealing with complex percentages, break them down into their components. For example, $17.5\% = 10\% + 5\% + 2.5\%$.

Graphs and Tables

When you're working with graphs and tables, it's important that you take a few seconds to check the following before actually extracting data from it.

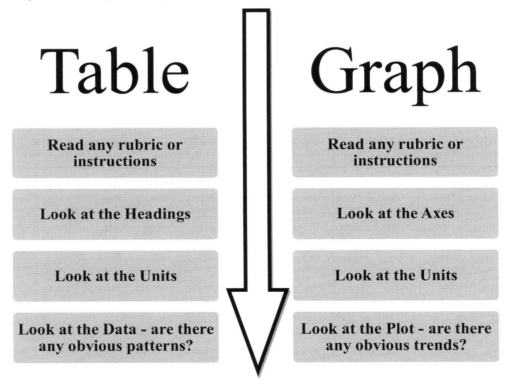

Get into the habit of doing this whenever you are faced with data and you'll find it much easier to approach these questions under time pressure.

Data Analysis Questions

Questions 25 to 27 are based on the following passage:

It has recently been questioned as to whether the recommended five fruit and vegetables a day is sufficient or if it would be more beneficial to eat 7 fruit and vegetable portions each day. A study at UCL looked at the fruit and vegetables eating habits of 65,000 people in England. Analysis of the data showed that eating more portions was beneficial and vegetables seemed to have a greater protective effect than fruit. The study however did not distinguish whether vegetables themselves have a greater protective effect, or whether these people tend to eat an overall healthier diet. A meta-analysis carried out by researchers across the world complied data from 16 studies which encompassed over 800,000 participants, of whom 56,423 had died.

They found a decline in death of around 5% from all causes for each additional portion of fruit or vegetables eaten, however they recorded no further decline for people who ate over 5 portions. Rates of cardiovascular disease, heart disease or stroke, were shown to decline 4% for each portion up to five, whereas the number of portions of fruit and vegetables eaten seemed to have little impact on cancer rates. The data from these studies points in a similar direction, that eating as much fruit and vegetables a day is preferable, but that five portions is sufficient to have a significant impact on reduction in mortality. Further studies need to look into the slight discrepancies, particularly why the English study found vegetables more protective, and if any specific cancers may be affected by fruit and vegetables even if the general cancer rates more greatly depend on other lifestyle factors.

Question 25:

Which of the following statements is correct?

A. The UCL study found no additional reduction in mortality in those who eat 7 rather than 5 portions of fruit and vegetables a day.

B. People who eat more fruit and vegetables are assumed to have an overall healthier diet which is what gives them the beneficial effect.

C. The meta analysis found fruit and vegetables are more protective against cancer than cardiovascular disease

D. The English study showed fruit had more protective effects than vegetables.

E. The meta-analysis found no additional reduction in mortality in those who eat 7 rather than 5 portions of fruit and vegetables a day.

F. The meta-analysis suggests people who eat 7 portions would have a 10% lower risk of death from any cause than those who eat 5 portions.

G. Fruit and vegetables are not protective against any specific cancers.

Question 26:

If rates of death were found to be 1% lower in the UCL study than the meta-analysis, approximately how many people died in the UCL study?

A. 3,000   B. 3,200   C. 3,900   D. 4,550   E. 5,200

Question 27:

Which statement does the article **MOST** agree with?

A. Eating more fruit and vegetables does not particularly lower the risk of any specific cancers.

B. The UCL research suggests that the guideline should be 7 fruit and vegetables a day for England.

C. The results found by the UCL study and the meta-analysis were contradictory.

D. Many don't eat enough vegetables due to cost and taste.

E. Fruit and vegetables are only protective against cardiovascular disease.

F. The UCL study and meta-analysis use a similar sample of participants.

G. People should aim to eat 7 portions of fruit and vegetables a day.

Questions 28-30 relate to the following table regarding average alcohol consumption in 2010.

Country	Total	Recorded Consumption	Unrecorded consumption	Beer (%)	Wine (%)	Spirits (%)	Other (%)	2020 Consumption Projection
Belarus		14.4	3.2	17.3	5.2	46.6	30.9	17.1
Lithuania	15.4	12.9	2.5		7.8	34.1	11.6	16.2
Andorra	13.8		1.4	34.6		20.1	0	9.1
Grenada	12.5	11.9	0.7	29.3	4.3		0.2	10.4
Czech Republic	13	11.8	1.2	53.5	20.5	26	0	14.1
France	12.2	11.8		18.8	56.4	23.1	1.7	11.6
Russia		11.5	3.6	37.6	11.4	51	0	14.5
Ireland	11.9	11.4	0.5	48.1	26.1	18.7	7.7	10.9

NB: Some data is missing.

Question 28:

Which of the following countries had the highest total beer and wine consumption for 2010?

A. Belarus
B. Lithuania
C. Ireland
D. France
E. Andorra

Question 29:

Which country has the greatest difference for spirit consumption in 2010 and 2020 projection, assuming percentages stay the same?

A. Russia
B. Belarus
C. Lithuania
D. Grenada
E. Ireland

Question 30:

It was later found that some of the percentages of types of alcohol consumed had been mixed up. If the actual amount of beer consumed by each person in the Czech Republic was on average 4.9L, which country were the percentage figures mixed up with?

A. Lithuania
B. Grenada
C. Russia
D. France
E. Ireland
F. Belarus
G. Andorra

Questions 31-33 are based on the following information:

The table below shows the incidence of 6 different types of cancer in Australia:

	Prostate	Lung	Bowel	Bladder	Breast	Uterus
Men	40,000	25,000	20,000	8,000	1,000	0
Women	0	20,000	18,000	4,000	50,000	9,000

Question 31:

Supposing there are 10 million men and 10 million women in Australia, how many percentage points higher is the incidence of cancer amongst women than amongst men?

A. 0.007 %   B. 0.07 %   C. 0.093 %   D. 0.7 %   E. 0.93 %

Question 32:

Now suppose there are 11.5 million men and 10 million women in Australia. Assuming all men are equally likely to get each type of cancer and all women are equally likely to get each type of cancer, how many of the types of cancer are you more likely to develop if you are a man than if you are a woman?

A. 1   B. 2   C. 3   D. 4

Question 33:

Suppose that prostate, bladder and breast cancer patients visit hospital 1 time during the first month of 2015 and patients for all other cancers visit hospital 2 times during the first month of 2015. 10% of cancer patients in Australia are in Sydney, and patients in Sydney are not more or less likely to have certain types of cancer than other patients.

How many hospital visits are made by patients in Sydney with these 6 cancers during the first month of 2015?

A. 10,300   C. 19,500   E. 195,000
B. 18,400   D. 28,700   F. 287,000

Question 34:

Which of the graphs correctly represents the combined proportion of men versus women with bladder cancer?

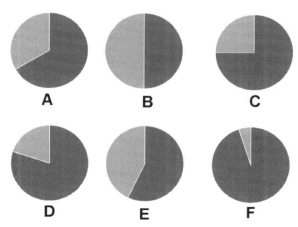

Questions 35 – 37 are based on the following information:

Units of alcohol are calculated by multiplying the alcohol percentage by the volume of liquid in litres, for example a 0.75 L bottle of wine which is 12% alcohol contains 9 units. 1 pint = 570 ml.

	Volume in bottle/barrel	Standard drinks per bottle/barrel	Percentage
Vodka	1250 ml	50	40%
Beer	10 pints	11.4	3%
Cocktail	750 ml	3	8%
Wine	750 ml	3.75	12.5%

Question 35:

Which standard drink has the most units of alcohol in?

A. Vodka
B. Beer
C. Cocktail
D. Wine

Question 36:

Some guidance suggests the recommended maximum number of units of alcohol per week for women is 14. In a week, Hannah drinks 4 standard drinks of wine, 3 standard drinks of beer, 2 standard cocktails and 5 standard vodkas. This guidance states the recommended maximum number of units per week for men is 21. In a week, Mark drinks 2 standard drinks of wine, 6 standard drinks of beer, 3 standard cocktails and 10 standard vodkas.

Who has exceeded their recommended maximum number of units by more and by how many units more have they exceeded it by than the other person?

A. Hannah, by 1 unit
B. Hannah, by 0.5 units
C. Both by the same
D. Mark, by 0.5 units
E. Mark, by 1 unit

Question 37:

How many different combinations of drinks that total 4 units are there (the same combination in a different order doesn't count).

A. 2
B. 3
C. 4
D. 5
E. 6

Questions 38-40 relate to the table below which shows information about Greentown's population:

	Female	Male	Total
Under 20	1,930		
20-39	1,960	3,760	5,720
40-59		4,130	
60 and over	2,350	2,250	4,600
Total	11,430	12,890	24,320

Question 38:

How many males under 20 are there in Greentown?

A. 2,650
B. 2,700
C. 2,730
D. 2,750
E. 2,850

Question 39:

How many females aged 40-59 are there in Greentown?

A. Between 3,000 and 4,000
B. Between 4,000 and 5,000
C. Between 5,000 and 6,000
D. Between 6,000 and 7,000

Question 40:

Which is the approximate ratio of females:males in the age group that has the highest ratio of males:females?

A. 1.4:1
B. 1.9:1
C. 1:1.9
D. 1:1.4

For hundred more BMAT practice questions check the *Ultimate BMAT Guide*– flick to the back to get a free copy.

SECTION 2: Scientific Knowledge and Application

Section 2 is undoubtedly the most time-pressured section of the BMAT. This section tests GCSE biology, chemistry, physics and maths. You have to answer 27 questions in 30 minutes. The questions can be quite difficult and it's easy to get bogged down. However, it's also the section in which you can improve the most quickly in so it's well worth spending time on it.

Although the vast majority of questions in section 2 aren't particularly difficult, the intense time pressure of having to do one question every minute makes this section the hardest in the BMAT. As with section 1, the trick is to identify and do the easy questions whilst leaving the hard ones for the end. In general, the biology and chemistry questions in the BMAT require the least amount of time per question whilst the maths and physics are more time-draining as they usually consist of multi-step calculations.

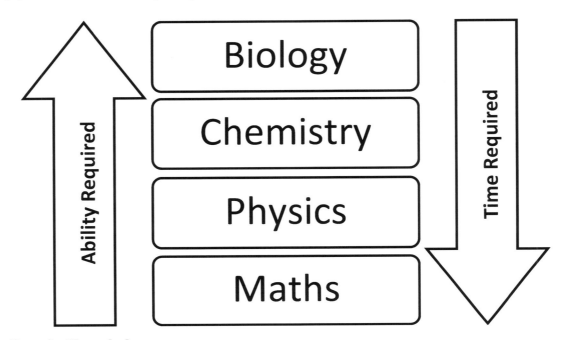

Gaps in Knowledge

The BMAT only tests GCSE level knowledge. However, there is a large variation in content between the GCSE exam boards meaning that you may not have covered some topics that are examinable. This is more likely if you didn't carry on with Biology or physics to AS level (e.g. Newtonian mechanics and parallel circuits in physics; hormones and stem cells in biology). If you fall into this category, you are highly advised to go through the BMAT Specification on the BMAT website here www.admissionstestingservice.org/bmat.

The questions in this book will help highlight any particular areas of weakness or gaps in your knowledge that you may have. Upon discovering these, make sure you take some time to revise these topics before carrying on – there is little to be gained by attempting section 2 questions with huge gaps in your knowledge.

Maths

Being confident with maths is extremely important for section 2. Many students find that improving their numerical and algebraic skills usually results in big improvements in their section 1 and 2 scores. Remember that maths in section 2 not only comes up in the maths question but also in physics (manipulating equations and standard form) and chemistry (mass calculations). So if you find yourself consistently running out of time in section 2, spending a few hours on brushing up your basic maths skills may do wonders for you.

SECTION 2: Biology

Thankfully, the biology questions tend to be fairly straightforward and require the least amount of time. You should be able to do the majority of these within the 60 second limit (often far less). This means that you should be aiming to make up time in these questions. In the majority of cases – you'll either know the answer or not i.e. they test advanced recall so the trick is to ensure that there are no obvious gaps in your knowledge.

Before going onto to do the practice questions in this book, ensure you are comfortable with the following commonly tested topics:

➢ Structure of animal, plant and bacterial cells

➢ Osmosis, Diffusion and Active Transport

➢ Cell Division (mitosis + meiosis)

➢ Family pedigrees and Inheritance

➢ DNA structure and replication

➢ Gene Technology & Stem Cells

➢ Enzymes – Function, mechanism and examples of digestive enzymes

➢ Aerobic and Anaerobic Respiration

➢ The central vs. peripheral nervous system

➢ The respiratory cycle including movement of ribs and diaphragm

➢ The Cardiac Cycle

➢ Hormones

➢ Basic immunology

➢ Food chains and food webs

➢ The carbon and nitrogen cycles

Top tip! If you find yourself getting less than 50% of biology questions correct in this book, make sure you revisit the syllabus before attempting more questions as this is the best way to maximise your efficiency. In general, there is no reason why you shouldn't be able to get the vast majority of biology questions correct (and in well under 60 seconds) with sufficient practice.

Biology Questions

Question 41:
In relation to the human genome, which of the following are correct?

1. The DNA genome is coded by 4 different bases.
2. The sugar backbone of the DNA strand is formed of glucose.
3. DNA is found in the nucleus of bacteria.

A. 1 only
B. 2 only
C. 3 only
D. 1 and 2
E. 1 and 3
F. 2 and 3
G. 1, 2 and 3

Question 42:
Animal cells contain organelles that take part in vital processes. Which of the following is true?

1. The majority of energy production by animal cells occurs in the mitochondria.
2. The cell wall protects the animal cell membrane from outside pressure differences.
3. The endoplasmic reticulum plays a role in protein synthesis.

A. 1 only
B. 2 only
C. 3 only
D. 1 and 2
E. 2 and 3
F. 1 and 3
G. 1, 2 and 3

Question 43:
With regards to animal mitochondria, which of the following is correct?

A. Mitochondria are not necessary for aerobic respiration.
B. Mitochondria are the sole cause of sperm cell movement.
C. The majority of DNA replication happens inside mitochondria.
D. Mitochondria are more abundant in fat cells than in skeletal muscle.
E. The majority of protein synthesis occurs in mitochondria.
F. Mitochondria are enveloped by a double membrane.

Question 44:
In relation to bacteria, which of the following is **FALSE**?

A. Bacteria always lead to disease.
B. Bacteria contain plasmid DNA.
C. Bacteria do not contain mitochondria.
D. Bacteria have a cell wall and a plasma membrane.
E. Some bacteria are susceptible to antibiotics.

Question 45:

In relation to bacterial replication, which of the following is correct?

A. Bacteria undergo sexual reproduction.
B. Bacteria have a nucleus.
C. Bacteria carry genetic information on circular plasmids.
D. Bacterial genomes are formed of RNA instead of DNA.
E. Bacteria require gametes to replicate.

Question 46:

Which of the following are correct regarding active transport?

A. ATP is necessary and sufficient for active transport.
B. ATP is not necessary but sufficient for active transport.
C. The relative concentrations of the material being transported have little impact on the rate of active transport.
D. Transport proteins are necessary and sufficient for active transport.
E. Active transport relies on transport proteins that are powered by an electrochemical gradient.

Question 47:

Concerning mammalian reproduction, which of the following is **FALSE**?

A. Fertilisation involves the fusion of two gametes.
B. Reproduction is sexual and the offspring display genetic variation.
C. Reproduction relies upon the exchange of genetic material.
D. Mammalian gametes are diploid cells produced via meiosis.
E. Embryonic growth requires carefully controlled mitosis.

Question 48:

Which of the following apply to Mendelian inheritance?

1. It only applies to plants.
2. It treats different traits as either dominant or recessive.
3. Heterozygotes have a 25% chance of expressing a recessive trait.

A. 1 only
B. 2 only
C. 3 only
D. 1 and 2
E. 1 and 3
F. 2 and 3
G. All of the above.

Question 49:

Which of the following statements are correct?

A. Hormones are secreted into the blood stream and act over long distances at specific target organs.
B. Hormones are substances that almost always cause muscles to contract.
C. Hormones have no impact on the nervous or enteric systems.
D. Hormones are always derived from food and never synthesised.
E. Hormones act rapidly to restore homeostasis.

Question 50:
With regard to neuronal signalling in the body, which of the following are true?

1. Neuronal transmission can be caused by both electrical and chemical stimulation.
2. Synapses ultimately result in the production of an electrical current for signal transduction.
3. All synapses in humans are electrical and unidirectional.

A. 1 only
B. 2 only
C. 3 only
D. 1 and 2
E. 1 and 3
F. 2 and 3
G. 1, 2 and 3

Question 51:
What is the **primary** reason that pH is controlled so tightly in humans?

A. To allow rapid protein synthesis.
B. To allow for effective digestion throughout the GI tract.
C. To ensure ions can function properly in neural signalling.
D. To prevent changes in electrical charge in polypeptide chains.
E. To prevent changes in core body temperature.

Question 52:
Which of the following statements are correct regarding cell walls?

1. The cell wall confers protection against external environmental stimuli.
2. The cell wall is an evolutionary remnant and now has little functional significance in most bacteria.
3. The cell wall is made up primarily of glucose.

A. Only 1
B. Only 2
C. Only 3
D. 1 and 2
E. 2 and 3
F. 1 and 3
G. 1, 2 and 3

For a hundred more BMAT practice questions check the *Ultimate BMAT Guide*– flick to the back to get a free copy.

SECTION 2: Chemistry

Most students don't struggle with BMAT chemistry as they'll be studying it at A2. However, there are certain questions that even good students tend to struggle with under time pressure e.g. balancing equations and mass calculations. It is essential that you're able to do these quickly as they take up by far the most time in the chemistry questions.

Balancing Equations

For some reason, most students are rarely shown how to formally balance equations – including those studying it at A-level. Balancing equations intuitively or via trial and error will only get you so far in the BMAT as the equations you'll have to work with will be fairly complex. To avoid wasting valuable time, it is essential you learn a method that will allow you to solve these in less than 60 seconds on a consistent basis. The method shown below is the simplest way and requires you to be able to do quick mental arithmetic (which is something you should be aiming for anyway). The easiest way to do learn it is through an example:

The following equation shows the reaction between Iodic acid, hydrochloric acid and copper Iodide:

$$\textbf{a } HIO_3 + \textbf{b } CuI_2 + \textbf{c } HCl \rightarrow \textbf{d } CuCl_3 + \textbf{e } ICl + \textbf{f } H_2O$$

What values of **a**, **b**, **c**, **d**, **e** and **f** are needed in order to balance the equation?

	a	b	c	d	e	f
A	5	4	25	4	13	15
B	5	4	20	4	8	15
C	5	6	20	6	8	15
D	2	8	10	8	8	15
E	6	8	24	10	16	15
F	6	10	22	10	16	15

Step 1: Pick an element and see how many atoms there are on the left and right sides.
Step 2: Form an equation to represent this. For Cu: b = d
Step 3: See if any of the answer options don't satisfy b=d. In this case, for option E, b is 8 and d is 10. This allows us to eliminate option E.

Once you've eliminated as many options as possible, go back to step 1 and pick another element.
For Hydrogen (H): a + c = 2f

Then see if any of the answer options don't satisfy a + c = 2f.
➢ Option A: 5 + 25 is equal to 2 x 15
➢ Option B: 5 + 20 is not equal to 2 x 15
➢ Option C: 5 + 20 is not equal to 2 x 15
➢ Option D: 2 + 10 is not equal to 2 x 15

This allows us to eliminate option B, C and D. E has already been eliminated. Thus, the only solution possible is A. This method works best when you get given a table above as this allows you to quickly eliminate options. However, it is still a viable method even if you don't get this information.

Chemistry Questions

Question 53:

Which of the following most accurately defines an isotope?

A. An isotope is an atom of an element that has the same number of protons in the nucleus but a different number of neutrons orbiting the nucleus.
B. An isotope is an atom of an element that has the same number of neutrons in the nucleus but a different number of protons orbiting the nucleus.
C. An isotope is any atom of an element that can be split to produce nuclear energy.
D. An isotope is an atom of an element that has the same number of protons in the nucleus but a different number of neutrons in the nucleus.
E. An isotope is an atom of an element that has the same number of protons in the nucleus but a different number of electrons orbiting it.

Question 54:

Which of the following is an example of a displacement reaction?

1. $Fe + SnSO4 \rightarrow FeSO_4 + Sn$
2. $Cl_2 + 2KBr \rightarrow Br_2 + 2KCl$
3. $H_2SO_4 + Mg \rightarrow MgSO_4 + H_2$
4. $Pb(NO_3)_2 + 2NaCl \rightarrow PbCl_2 + 2NaNO_3$

A. 1 only
B. 1 and 2 only
C. 2 and 3 only
D. 3 and 4 only
E. 1, 2 and 3 only
F. 2, 3 and 4 only

Question 55:

What values of **a**, **b** and **c** are needed to balance the equation below?

$$aCa(OH)_2 + bH_3PO_4 \rightarrow Ca_3(PO_4)_2 + cH_2O$$

A. a = 3 b = 2 c = 6
B. a = 2 b = 2 c = 4
C. a = 3 b = 2 c = 1
D. a = 1 b = 2 c = 3
E. a = 4 b = 2 c = 6
F. a = 3 b = 2 c = 4

Question 56:

What values of **s**, **t** and **u** are needed to balance the equation below?

$$sAgNO_3 + tK_3PO_4 \rightarrow 3Ag_3PO_4 + uKNO_3$$

A. s = 9 t = 3 u = 9
B. s = 6 t = 3 u = 9
C. s = 9 t = 3 u = 6
D. s = 9 t = 6 u = 9
E. s = 3 t = 3 u = 9
F. s = 9 t = 3 u = 3

Question 57:
Which of the following statements are true with regard to displacement?

1. A less reactive halogen can displace a more reactive halogen.
2. Chlorine cannot displace bromine or iodine from an aqueous solution of its salts.
3. Bromine can displace iodine because of the trend of reactivity.
4. Fluorine can displace chlorine as it is higher up the group.
5. Lithium can displace francium as it is higher up the group.

A. 3 only
B. 5 only
C. 1 and 2 only

D. 3 and 4 only
E. 2 , 3 and 5 only
F. 3, 4 and 5 only

Question 58:
What mass of magnesium oxide is produced when 75g of magnesium is burned in excess oxygen?
Relative Atomic Masses: $Mg = 24$, $O = 16$

A. 80g   B. 100g   C. 125g   D. 145g   E. 175g   F. 225g

Question 59:
Hydrogen can combine with hydroxide ions to produce water. Which process is involved in this?

A. Hydration
B. Oxidation

C. Reduction
D. Dehydration

E. Evaporation
F. Precipitation

Question 60:
Which of the following statements about Ammonia are correct?

1. It has a formula of NH_3.
2. Nitrogen contributes 82% to its mass.
3. It can be broken down again into nitrogen and hydrogen.
4. It is covalently bonded.
5. It is used to make fertilisers.

A. 1 and 2 only
B. 1 and 4 only
C. 1, 2 and 3 only

D. 1, 2 and 5 only
E. 3, 4 and 5 only
F. 1, 2, 3, 4 and 5

Question 61:
What colour will a universal indicator change to in a solution of milk and lipase?

A. From green to orange.
B. From red to green.
C. From purple to green.

D. From purple to orange.
E. From yellow to purple.
F. From purple to red.

Question 62:
Vitamin C [$C_6H_8O_6$] can be artificially synthesised from glucose [$C_6H_{12}O_6$]. What type of reaction is this likely to be?

A. Dehydration
B. Hydration

C. Oxidation
D. Reduction

E. Displacement
F. Evaporation

For hundred more BMAT practice questions check the *Ultimate BMAT Guide*– flick to the back to get a free copy.

SECTION 2: Physics

If you haven't done physics at AS then you'll have to ensure that you are confident with commonly examined topics like Newtonian mechanics, electrical circuits and radioactive decay. With certain specifications you may not have covered these at GCSE level.

The first step to improving in this section is to memorise by rote all the equations listed on the next page.

The majority of the physics questions involve a fair bit of maths – this means you need to be comfortable with converting between units and also powers of 10. **Most questions require two step calculations**. Consider the example:

A metal ball is released from the roof a 20 metre building. Assuming air resistance equals is negligible; calculate the velocity at which the ball hits the ground. [$g = 10ms^{-2}$]

A. $5 \ ms^{-1}$
B. $10 \ ms^{-1}$
C. $15 \ ms^{-1}$
D. $20 \ ms^{-1}$
E. $25 \ ms^{-1}$

When the ball hits the ground, all of its gravitational potential energy has been converted to kinetic energy. Thus, $E_p = E_k$:

$$mg\Delta h = \frac{mv^2}{2}$$

Thus, $v = \sqrt{2gh} = \sqrt{2 \times 10 \times 20}$

$= \sqrt{400} = 20ms^{-1}$

Here, you were required to not only recall two equations but apply and rearrange them very quickly to get the answer; all in under 60 seconds. Thus, it is easy to understand why the physics questions are generally much harder than the biology and chemistry ones.

Note that if you were comfortable with basic Newtonian mechanics, you could have also solved this using a single SUVAT equation: $v^2 = u^2 + 2as$

$v = \sqrt{2 \times 10 \times 20} = 20ms^{-1}$

This is why you're **strongly advised to learn the 'suvat' equations** on the next page even if they're technically not on the syllabus.

SI Units

Remember that in order to get the correct answer you must always work in SI units i.e. do your calculations in terms of metres (not centimetres) and kilograms (not grams), etc.

> ***Top tip!*** Knowing SI units is extremely useful because they allow you to **'work out' equations** if you ever forget them e.g. The units for density are kg/m^3. Since Kg is the SI unit for mass, and m^3 is represented by volume –the equation for density must be = Mass/Volume. This can also work the other way, for example we know that the unit for Pressure is Pascal (Pa). But based on the fact that Pressure = Force/Area, a Pascal must be equivalent to N/m^2. Some physics questions will test your ability to manipulate units like this, so it's important you are comfortable converting between them.

Formulae you MUST know:

Equations of Motion:

- $s = ut + 0.5at^2$
- $v = u + at$
- $a = (v-u)/t$
- $v^2 = u^2 + 2as$

Equations relating to Force:

- Force = mass x acceleration
- Force = Momentum/Time
- Pressure = Force / Area
- Moment of a Force = Force x Distance
- Work done = Force x Displacement

For objects in equilibrium:

- Sum of Clockwise moments = Sum of Anti-clockwise moments
- Sum of all resultant forces = 0

Equations relating to Energy:

- Kinetic Energy = $0.5\ mv^2$
- Δ in Gravitational Potential Energy = $mg\Delta h$
- Energy Efficiency = (Useful energy/ Total energy) x 100%

Equations relating to Power:

- Power = Work done / time
- Power = Energy transferred / time
- Power = Force x velocity

Electrical Equations:

- $Q = It$
- $V = IR$
- $P = IV = I^2R = V^2/R$
- V = Potential difference (V, Volts)

- R = Resistance (Ohms)
- P = Power (W, Watts)
- Q = Charge (C, Coulombs)
- t = Time (s, seconds)

For Transformers: $\dfrac{V_p}{V_s} = \dfrac{n_p}{n_s}$ where:

- V: Potential difference
- n: Number of turns
- p: Primary
- s: Secondary

Other:

- Weight = mass x g
- Density = Mass / Volume
- Momentum = Mass x Velocity
- $g = 9.81\ ms^{-2}$ (unless otherwise stated)

Factor	Text	Symbol
10^{12}	Tera	T
10^{9}	Giga	G
10^{6}	Mega	M
10^{3}	Kilo	k
10^{2}	Hecto	h
10^{-1}	Deci	d
10^{-2}	Centi	c
10^{-3}	Milli	m
10^{-6}	Micro	μ
10^{-9}	Nano	n
10^{-12}	Pico	p

Physics Questions

Question 63:

Which of the following statements are **FALSE**?

A. Electromagnetic waves cause things to heat up.
B. X-rays and gamma rays can knock electrons out of their orbits.
C. Loud sounds can make objects vibrate.
D. Wave power can be used to generate electricity.
E. Since waves carry energy away, the source of a wave loses energy.
F. The amplitude of a wave determines its mass.

Question 64:

A spacecraft is analysing a newly discovered exoplanet. A rock of unknown mass falls on the planet from a height of 30 m. Given that $g = 5.4$ ms^{-2} on the planet, calculate the speed of the rock when it hits the ground and the time it took to fall.

	Speed (ms^{-1})	Time (s)
A	18	3.3
B	18	3.1
C	12	3.3
D	10	3.7
E	9	2.3
F	1	0.3

Question 65:

A canoe floating on the sea rises and falls 7 times in 49 seconds. The waves pass it at a speed of 5 ms^{-1}. How long are the waves?

A. 12 m   B. 22 m   C. 25 m   D. 35 m   E. 57 m   F. 75 m

Question 66:

Miss Orrell lifts her 37.5 kg bike for a distance of 1.3 m in 5 s. The acceleration of free fall is 10 ms^{-2}. What is the average power that she develops?

A. 9.8 W   C. 57.9 W   E. 97.5W
B. 12.9 W   D. 79.5 W   F. 98.0 W

Question 67:

A truck accelerates at 5.6 ms^{-2} from rest for 8 seconds. Calculate the final speed and the distance travelled in 8 seconds.

	Final Speed (ms^{-1})	Distance (m)
A	40.8	119.2
B	40.8	129.6
C	42.8	179.2
D	44.1	139.2
E	44.1	179.7
F	44.2	129.2
G	44.8	179.2
H	44.8	179.7

Question 68:

Which of the following statements is true when a sky diver jumps out of a plane?

A. The sky diver leaves the plane and will accelerate until the air resistance is greater than their weight.
B. The sky diver leaves the plane and will accelerate until the air resistance is less than their weight.
C. The sky diver leaves the plane and will accelerate until the air resistance equals their weight.
D. The sky diver leaves the plane and will accelerate until the air resistance equals their weight squared.
E. The sky diver will travel at a constant velocity after leaving the plane.

Question 69:

A 100 g apple falls on Isaac's head from a height of 20 m. Calculate the apple's momentum before the point of impact. Take $g = 10$ ms^{-2}

A. 0.1 kgms^{-1}
B. 0.2 kgms^{-1}
C. 1 kgms^{-1}
D. 2 kgms^{-1}
E. 10 kgms^{-1}
F. 20 kgms^{-1}

Question 70:

Which of the following do all electromagnetic waves all have in common?

1. They can travel through a vacuum.
2. They can be reflected.
3. They are the same length.
4. They have the same amount of energy.
5. They can be polarised.

A. 1, 2 and 3 only
B. 1, 2, 3 and 4 only
C. 4 and 5 only
D. 3 and 4 only
E. 1, 2 and 5 only
F. 1 and 5 only

Question 71:

A battery with an internal resistance of 0.8 Ω and e.m.f of 36 V is used to power a drill with resistance 1 Ω. What is the current in the circuit when the drill is connected to the power supply?

A. 5 A   B. 10 A   C. 15 A   D. 20 A   E. 25 A   F. 30 A

Question 72:

Officer Bailey throws a 20 g dart at a speed of 100 ms^{-1}. It strikes the dartboard and is brought to rest in 10 milliseconds. Calculate the average force exerted on the dart by the dartboard.

A. 0.2 N
B. 2 N
C. 20 N
D. 200 N
E. 2,000 N
F. 20,000 N

Question 73:

Professor Huang lifts a 50 kg bag through a distance of 0.7 m in 3 s. What average power does she develop to 3 significant figures? Take $g = 10$ms^{-2}

A. 112 W
B. 113 W
C. 114 W
D. 115 W
E. 116 W
F. 117 W

For hundred more BMAT practice questions check the *Ultimate BMAT Guide*– flick to the back to get a free copy.

SECTION 2: Maths

BMAT Maths questions are designed to be time draining- if you find yourself consistently not finishing, it might be worth leaving the maths (and probably physics) questions until the very end.

Good students sometimes have a habit of making easy questions difficult; remember that the BMAT only tests GCSE level knowledge so you are not expected to know or use calculus or trigonometry in any part of the exam.

Formulas you **MUST** know:

2D Shapes		3D Shapes		
	Area		Surface Area	Volume
Circle	πr^2	Cuboid	Sum of all 6 faces	Length x width x height
Parallelogram	Base x Vertical height	Cylinder	$2\pi r^2 + 2\pi rl$	πr^2 x l
Trapezium	0.5 x h x (a+b)	Cone	$\pi r^2 + \pi rl$	πr^2 x (h/3)
Triangle	0.5 x base x height	Sphere	$4\pi r^2$	$(4/3)\pi r^3$

Even good students who are studying maths at A2 can struggle with certain BMAT maths topics because they're usually glossed over at school. These include:

Quadratic Formula

The solutions for a quadratic equation in the form $ax^2 + bx + c = 0$ are given by: $x = \frac{-b \pm \sqrt{b^2 - 4ac}}{2a}$

Remember that you can also use the discriminant to quickly see if a quadratic equation has any solutions:

$$If\ b^2 - 4ac < 0: No\ solutions$$
$$If\ b^2 - 4ac = 0: One\ solution$$
$$If\ b^2 - 4ac > 2: Two\ solutions$$

Completing the Square

If a quadratic equation cannot be factorised easily and is in the format $ax^2 + bx + c = 0$ then you can rearrange it into the form $a\left(x + \frac{b}{2a}\right)^2 + \left[c - \frac{b^2}{4a}\right] = 0$

This looks more complicated than it is – remember that in the BMAT, you're extremely unlikely to get quadratic equations where $a > 1$ and the equation doesn't have any easy factors. This gives you an easier equation:

$\left(x + \frac{b}{2}\right)^2 + \left[c - \frac{b^2}{4}\right] = 0$ and is best understood with an example.

Consider: $x^2 + 6x + 10 = 0$

This equation cannot be factorised easily but note that: $x^2 + 6x - 10 = (x + 3)^2 - 19 = 0$

Therefore, $x = -3 \pm \sqrt{19}$. Completing the square is an important skill – make sure you're comfortable with it.

Difference between 2 Squares

If you are asked to simplify expressions and find that there are no common factors but it involves square numbers – you might be able to factorise by using the 'difference between two squares'.

For example, $x^2 - 25$ can also be expressed as $(x + 5)(x - 5)$.

Maths Questions

Question 74:

Robert has a box of building blocks. The box contains 8 yellow blocks and 12 red blocks. He picks three blocks from the box and stacks them up high. Calculate the probability that he stacks two red building blocks and one yellow building block, in **any** order.

A. $\frac{8}{20}$　　　B. $\frac{44}{95}$　　　C. $\frac{11}{18}$　　　D. $\frac{8}{19}$　　　E. $\frac{12}{20}$　　　F. $\frac{35}{60}$

Question 75:

Solve $\frac{3x+5}{5} + \frac{2x-2}{3} = 18$

A. 12.11　　　B. 13.49　　　C. 13.95　　　D. 14.2　　　E. 19　　　F. 265

Question 76:

Solve $3x^2 + 11x - 20 = 0$

A. 0.75 and $-\frac{4}{3}$　　　　C. -5 and $\frac{4}{3}$　　　　E. 12 only

B. -0.75 and $\frac{4}{3}$　　　　D. 5 and $\frac{4}{3}$　　　　F. -12 only

Question 77:

Express $\frac{5}{x+2} + \frac{3}{x-4}$ as a single fraction.

A. $\frac{15x-120}{(x+2)(x-4)}$　　　　C. $\frac{8x-14}{(x+2)(x-4)}$　　　　E. 24

B. $\frac{8x-26}{(x+2)(x-4)}$　　　　D. $\frac{15}{8x}$　　　　F. $\frac{8x-14}{x^2-8}$

Question 78:

The value of p is directly proportional to the cube root of q. When p = 12, q = 27. Find the value of q when p = 24.

A. 32　　　B. 64　　　C. 124　　　D. 128　　　E. 216　　　F. 1728

Question 79:

Write 72^2 as a product of its prime factors.

A. $2^6 \times 3^4$　　　　C. $2^4 \times 3^4$　　　　E. $2^6 \times 3$

B. $2^6 \times 3^5$　　　　D. 2×3^3　　　　F. $2^3 \times 3^2$

Question 80:

Calculate: $\dfrac{2.302 \times 10^5 + 2.302 \times 10^2}{1.151 \times 10^{10}}$

A. 0.0000202	C. 0.00002002	E. 0.000002002
B. 0.00020002	D. 0.00000002	F. 0.000002002

Question 81:

Given that $y^2 + \mathbf{a}y + \mathbf{b} = (y + 2)^2 - 5$, find the values of **a** and **b**.

	a	**b**
A	-1	4
B	1	9
C	-1	-9
D	-9	1
E	4	-1
F	4	1

Question 82:

Express $\dfrac{4}{5} + \dfrac{m-2n}{m+4n}$ as a single fraction in its simplest form:

A. $\dfrac{6m+6n}{5(m+4n)}$	C. $\dfrac{20m+6n}{5(m+4n)}$	E. $\dfrac{3(3m+2n)}{5(m+4n)}$
B. $\dfrac{9m+26n}{5(m+4n)}$	D. $\dfrac{3m+9n}{5(m+4n)}$	F. $\dfrac{6m+6n}{3(m+4n)}$

Question 83:

A is inversely proportional to the square root of B. When A = 4, B = 25.

Calculate the value of A when B = 16.

A. 0.8   B. 4   C. 5   D. 6   E. 10   F. 20

Question 84:

S, T, U and V are points on the circumference of a circle, and O is the centre of the circle.

Given that angle SVU = 89°, calculate the size of the smaller angle SOU.

A. 89°   B. 91°   C. 102°   D. 178°   E. 182°   F. 212°

Question 85:

Open cylinder A has a surface area of 8π cm^2 and a volume of 2π cm^3. Open cylinder B is an enlargement of A and has a surface area of 32π cm^2. Calculate the volume of cylinder B.

A. 2π cm^3	C. 10π cm^3	E. 16π cm^3
B. 8π cm^3	D. 14π cm^3	F. 32π cm^3

Section 3: Writing Task

In section 3, you have to write a one A4 page essay on one of four essay titles. Whilst different questions will inevitably demand differing levels of comprehension and knowledge, it is important to realise that one of the major skills being tested is actually your ability to construct a logical and coherent argument- and to convey it to the lay-reader.

Section 3 of the BMAT is frequently neglected by lots of students, who choose to spend their time on sections 1 & 2 instead. However, it has the highest returns per hour of work out of all three sections so is well worth putting time into.

The aim of section 3 is not to write as much as you can. Rather, the examiner is looking for you to make interesting and well supported points, and tie everything neatly together for a strong conclusion. Make sure you're writing critically and concisely; not rambling on. **Irrelevant material can actually lower your score.** You only get one side of A4 for your BMAT essay, so make it count!

Essay Structure

Most BMAT essays consist of 3 parts:

1) Explain what a quote or a statement means.
2) Argue for or against the statement.
3) Ask you "to what extent" you agree with the statement.

Number 1 should be the smallest portion of the essay (no more than 4 lines) and be used to provide a smooth segue into the rather more demanding "argue for/against" part of the question. This main part requires a firm grasp of the concept being discussed and the ability to strengthen and support the argument with a wide variety of examples from multiple fields. This section should give a balanced approach to the question, exploring **at least two distinct ideas**. Supporting evidence should be provided throughout the essay, with examples referred to when possible.

The final part effectively asks for your personal opinion and is a chance for you to shine- be brave and make an **innovative yet firmly grounded conclusion** for an exquisite mark. The conclusion should bring together all sides of the argument, in order to reach a clear and concise answer to the question. There should be an obvious logical structure to the essay, which reflects careful planning and preparation.

Paragraphs

Paragraphs are an important formatting tool which show that you have thought through your arguments and are able to structure your ideas clearly. A new paragraph should be used every time a new idea is introduced. There is no single correct way to arrange paragraphs, but it's important that each paragraph flows smoothly from the last. A slick, interconnected essay shows that you have the ability to communicate and organise your ideas effectively.

Given that you only have a limit of one A4 page to write in – **you shouldn't have more than 5 paragraphs** (use indents to show paragraphs – don't leave empty lines!). In general, 2 of these 5 will be taken up by the introduction and conclusion respectively.

Remember- the emphasis should remain on quality and not quantity. An essay with fewer paragraphs, but with well-developed ideas, is much more effective than a number of short, unsubstantial paragraphs that fail to fully grasp the question at hand.

Approaching the Essay

Section 3 can be broken down into 3 components; selecting your essay title, planning and writing it.

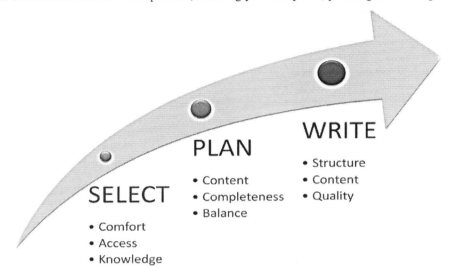

Most students think that the "writing" component is most important. This is simply not true.

The vast **majority of problems are caused by a lack of planning and essay selection**- usually because students just want to get writing as they are worried about finishing on time. Thirty minutes is long enough to be able to plan your essay well and *still* have time to write it so don't feel pressured to immediately start writing.

Step 1: Selecting

Selecting your essay is crucial- make sure you're comfortable with the topic and ensure you understand the actual question- it sounds silly but about 25% of essays that we mark score poorly because they don't actually answer the question!

Take two minutes to read all the questions. Whilst one essay might originally seem the easiest, if you haven't thought through it you might quickly find yourself running out of ideas. Likewise, a seemingly difficult essay might actually offer you a good opportunity to make interesting points.

Use this time to carefully select which question you will answer by gauging how accessible and comfortable you are with it given your background knowledge. Section 3, however, is not a test of knowledge but rather a test of how well you are able to argue.

It's surprisingly easy to change a question into something similar, but with a different meaning. Thus, you may end up answering a completely different essay title. Once you've decided which question you're going to do, read it very carefully through a few times to make sure you fully understand it. Answer all aspects of the question. Keep reading it as you answer to ensure you stay on track!

Step 2: Planning

Why should I plan my essay?
There are multiple reasons you should plan your essay for the first 5-10 minutes of section 3:

- As you don't have much space to write, make the most of it by writing a very well organised essay.
- It allows you to get all your thoughts ready before you put pen to paper.
- You'll write faster once you have a plan.
- You run the risk of missing the point of the essay or only answering part of it if you don't plan adequately.

How much time should I plan for?

There is no set period of time that should be dedicated to planning, and everyone will dedicate a different length of time to the planning process. You should spend as long planning your essay as you require, but it is essential that you leave enough time to write the essay. As a rough guide, it is **worth spending about 5-10 minutes to plan** and the remaining time on writing the essay. However, this is not a strict rule, and you are advised to tailor your time management to suit your individual style.

How should I go about the planning process?

There are a variety of methods that can be employed in order to plan essays (e.g. bullet-points, mind-maps etc). If you don't already know what works best, it's a good idea to experiment with different methods.

Generally, the first step is to gather ideas relevant to the question, which will form the basic arguments around which the essay is to be built. You can then begin to structure your essay, including the way that points will be linked. At this stage it is worth considering the balance of your argument, and confirming that you have considered arguments from both sides of the debate. Once this general structure has been established, it is useful to consider any examples or real world information that may help to support your arguments. Finally, you can begin to assess the plan as a whole, and establish what your conclusion will be based on your arguments.

Step 3: Writing

Introduction

Why are introductions important?

An introduction provides tutors with their first opportunity to examine your work. The introduction is where first impressions are formed, and these can be extremely important in producing a convincing argument. A well-constructed introduction shows that you have really thought about the question, and can indicate the logical flow of arguments that is to come.

What should an introduction do?

A good introduction should **briefly explain the statement or quote** and give any relevant background information in a concise manner. However, don't fall into the trap of just repeating the statement in a different way. The introduction is the first opportunity to suggest an answer to the question posed- the main body is effectively your justification for this answer.

Main Body

How do I go about making a convincing point?

Each idea that you propose should be supported and justified, in order to build a convincing overall argument. A point can be solidified through a basic Point → Evidence → Evaluation process.

How do I achieve a logical flow between ideas?

One of the most effective ways of displaying a good understanding of the question is to keep a logical flow throughout your essay. This means linking points effectively between paragraphs, and creating a congruent train of thought for the examiner as the argument develops. A good way to generate this flow of ideas is to provide ongoing comparisons of arguments, and discussing whether points support or dispute one another.

Should I use examples?

In short – yes! Examples can help boost the validity of arguments, and can help display high quality writing skills. Examples can add a lot of weight to your argument and make an essay much more relevant to the reader. When using examples, you should ensure that they are relevant to the point being made, as they will not help to support an argument if they are not.

Some questions will provide more opportunities to include examples than others so don't worry if you aren't able to use as many examples as you would have liked. There is no set rule about how many examples should be included!

Conclusion

The conclusion provides an opportunity to emphasise the **overall sentiment of your essay** which readers can then take away. It should summarise what has been discussed during the main body and give a definitive answer to the question. Some students use the conclusion to **introduce a new idea that hasn't been discussed**. This can be an interesting addition to an essay, and can help make you stand out. However, it is by no means, a necessity. In fact, a well-organised, 'standard' conclusion is likely to be more effective than an adventurous but poorly executed one.

> ***Top tip!*** Remember that there is no single correct answer to these questions and you're not expected to be able to fit everything onto one page. Instead it's better to pick a few key points to focus on.

Common Mistakes

Ignoring the other side of the argument

Although you're normally required to support one side of the debate, it is important to **consider arguments against your judgement** in order to get the higher marks. A good way to do this is to propose an argument that might be used against you, and then to argue why it doesn't hold true or seem relevant. You may use the format: *"some may say that…but this doesn't seem to be important because…"* in order to dispel opposition arguments, whilst still displaying that you have considered them. For example, *"some may say that fox hunting shouldn't be banned because it is a tradition. However, witch hunting was also once a tradition – we must move on with the times".*

Answering the topic/Answering only part of the question

One of the most common mistakes is to only answer a part of the question whilst ignoring the rest of it as it's inaccessible. According to the official mark scheme, **in order to get a score of 3 or more, you must write "…an answer that addresses ALL aspects of the question".** This should be your minimum standard- anything else that you write should then point you towards achieving 4/5.

Long Introductions

Some students can start rambling and make introductions too long and unfocussed. Although background information about the topic can be useful, it is normally not necessary. Instead, the **emphasis should be placed on responding to the question**. Some students also just **rephrase the question** rather than actually explaining it. The examiner knows what the question is, and repeating it in the introduction is simply a waste of space in an essay where you are limited to just one A4 side.

Not including a Conclusion

An essay that lacks a conclusion is incomplete and can signal that the answer has not been considered carefully or that your organisation skills are lacking. **The conclusion should be a distinct paragraph** in its own right and not just a couple of rushed lines at the end of the essay.

Sitting on the Fence

Students sometimes don't reach a clear conclusion. You need to **ensure that you give a decisive answer to the question** and clearly explain how you've reached this judgement. Essays that do not come to a clear conclusion generally have a smaller impact and score lower.

Exceeding the one page limit

The page limit is there for a reason – don't exceed it under any circumstances as any material over the limit won't be marked and it will appear that you haven't read the instructions.

Not using all the available space

Remember that you only have one A4 side to write on so ensure you make the maximum use of the space available to you. Don't leave lines to show paragraphs – instead, you should use indents. Similarly, you should also use the top-most line in the response sheet and avoid crossing entire sentences out.

Annotated Essays

Example Essay 1:

"A dentist should never disclose dental information about his patients"
What does this statement mean? Argue to the contrary using examples to strengthen your response.
To what extent do you agree with this statement?

The statement suggests that one of a dentist's most vital qualities is maintaining confidentiality of a patient's dental record. This involves all dentists with various specialities in different work places such as the clinics.

Disclosing dental informations regarding to their patients by dentists is considered as an unacceptable act within the dental society. For Example, by informing unrelated people about the patient might result in the individual's most embarrassed health situation to be exposed. For example, suffering pain from their private parts and this may disgust other. This situation would inevitably upset the patient as their health privacy has been breached by others without consent leading to a sence of distrust towards dentists.

However, disclosing such matters to certain suitable people such as family and relatives may be crucial. For example, if the patient is the head of the family or the guardian to the children. As these individuals are in charge of leading and taking are of the family, they need to be able to perform mundane task (such as providing good support to the children) at their optimum. Also, to ensure that the members realise that the patient should not over exert him or herself despite their health conditions. Also, a sudden collapse will reduce shock when the family is to rush the patient back to the hospital knowing that the illness is related to the situation.

Overall, a patient's confidentially should not be disclosed without consent or any importance by all means. This is to respect their health privacy and to avoid any inconvenience within the dental society.

Examiner's Comments:

Introduction: The student appears to have an understanding of the topic but frequently makes statements that don't add much to the argument e.g. the second sentence of the first paragraph. The introduction would be better used to set up the counter arguments that will form the bulk of the main body.

Main Body: The first paragraph actually supports the statement and is therefore not actually answering the question. The example doesn't really add much either. The key issue that needed to be discussed was a dentist's duty to the patient – not about "disgusting others". The last sentence of the first paragraph is good and starts to address the question but it comes far too late.

The second paragraph is better but misses the key points of the essay that were necessary to discuss i.e. when can patient confidentiality be broken? Examples would include suspected terrorism, notifiable diseases and criminal activity, suicides etc.

Conclusion: The conclusion doesn't really address the counter-arguments for breaking confidentiality or give a balanced answer. It also contains confusing terminology e.g. one discloses confidential information, not 'confidentiality'. The sentence concerning "inconvenience within the dental community" is also somewhat ambiguous.

Language: The grammar hinders the points that the student is trying to make throughout the essay e.g. "suffering pain from their private parts and this may disgust other". There are also frequent spelling mistakes like "embarrased" and "sence" that reduce the fluency significantly.

Score: D2

Example Essay 2

"A dentist should never disclose dental information about his patients"
What does this statement mean? Argue to the contrary using examples to strengthen your response.
To what extent do you agree with this statement?

This statement is one of the duties set out by the General Dental Council for dentists to comply with, which is to respect patient's autonomy. It means that a dentist cannot share patient's dental information to other parties unless the patients, themselves, have granted permissions to do so.
The ethical principle of respecting patients' autonomy cannot be applied in all cases, as some cases require dentists to disclose dental information about patients. First, when it involves a criminal act that has been comitted by a patient, a dentist has to report to any appropriate authority, such as the police. This is because the patient may potentially cause even more harm to others and as dentists, they have to prevent that from happening. An example of a case would be if a patient has suffered from a gunshot wound and he had told his dentist that he had gotten it when he murdered someone.

Next, another incident when a dentist has no choice but to disclose patients' dental information is when it may affect the health of society and could potentially cause an epidemic. Such patients might have an infectious disease and do not wish to let other people know about it. For instance, there has been many cases in West Africa where people who have Ebola are afraid to let their neighbours or friends find out because they do not want to be stigmatised and ostrilised from the society. However, these patients could spread the disease and so a dentist must not withold the information. Last of all, if a patient is underage, then he/ she is still not competent enough to make her own decision. Therefore, any dental information must be shared with his/ her legal guardian.

Respecting patient's autonomy by never disclosing their information is also important because patients have the right to chose who gets to know about it. It is his own body. He is the only person who knows the consequences of sharing this sensitive information. In conclusion, I believe that never disclosing patients' dental information cannot be complied in every incident. Respecting their autonomy is important but we have to treat each case separately.

Examiner's Comments:

Introduction: The introduction is well written but could be improved by making it explicitly clear that confidentiality can be breached in certain circumstances. This would then set up the main body nicely as the student would then be able to go straight into giving examples.

Main Body: The first sentence of the second paragraph is well written but should have gone in the introduction. There is good breadth of argument with the important points being covered like a dentist's duty to prevent harm to others, 'public good' and the issue of 'capacity'. However, there is unnecessary padding that doesn't add much e.g. there was no need to expand on your example of infectious diseases. The extra space from avoiding this would have allowed the student to write about the fact that although confidentiality must sometimes be breached, a dentist has certain professional duties. For example, informing the patient both before and after and explaining why they have disclosed what they have to try and mitigate any damage to the dentist patient relationship. In this way public trust in dental professionals would be maintained.

Conclusion: The conclusion concisely summarises the arguments put forth in the main body and offers a nice resolution by saying that each case is different.

Language: Whilst it is clear that the student understands the question – there is some confusion as to the difference between "autonomy" and "confidentiality" (It's important to know the basics of dental ethics as they'll be helpful for the interview stage as well).Furthermore, there are minor spelling mistakes like "comitted" and "withold". In the conclusion, they also assume that patients are only male – "it is his own body" vs. "it is their own body".

Score: B3.5

Example Essay 3

"A dentist should never disclose dental information about his patients"
What does this statement mean? Argue to the contrary using examples to strengthen your response.
To what extent do you agree with this statement?

Confidentiality is a basic patient right. The patient provides information to the dentist not to be unnecessarily shared with others without their knowledge or permission. On this basis, the statement argues that a dentist should never reveal the dental data, such as results from tests or prescriptions given.

However, it can be argued that there are many circumstances whereby it is necesary to breach patient confidentiality and disclose dental information. More specifically, if the patient poses a threat to the public health, their dental situation should be disclosed immediately so that actions can be taken to prevent the spread. For instance, under the Public Health Act 1988, if a patient is suspected of communicable diseases such as tuberculosis, the dentist is required to inform the local health authorities immediately so that they can make precautions to protect the other citizens. In addition, a dentist should also disclose dental information if the patient has broken the law. For instance, the dentist should reveal dental data to the detectives and other relevant professionals if they request for it, to enable them to come to a conclusion of the case more quickly and accurately.

However, I agree with this statement to a large extent. After all, the patient should have the right over what happens to his dental documents and information. Revealing information about the patient unnecessarily will take this basic right away, and it is extremely unfair for the patient. Furthermore, this unprofessional decision may undermine the confidence between the patient and dentist. The patient may be less willing to reveal vital personal information to the dentist in the future, in fear that he might release this information as well. This would be extremely detrimental to the diagnoses and treatment for the patient or the dentist might not be able to gain sufficient information to make a more informed decision.

In conclusion, a dentist should never disclose dental information about his patients unless there are other external circumstances that oblige the dentist to do so. Breaking this confidentiality will cause the patient-dentist relationship to collapse, compromising the trust between them. However, in some cases, the decision to disclose is not that clear-cut; if a patient had sexually-transmitted infections, should the dentist disclose this information to his spouse? Such situations have to be decided on a case-by-case basis.

Examiner's Comments:

Introduction: This is a bold introduction that catches one's attention and gets straight to the point. The student however does make a rather generalised statement - "confidentiality is a basic patient right". A pedantic examiner could easily challenge this and thus, it's important to be careful with your wording.

Main Body: There is a good level of breadth and depth of argument here. However, there are again some generalising statements that are incorrect e.g. dentists don't need to break confidentiality if a patient has broken ANY law – just serious ones e.g. committed or intend to commit murder/terrorism and dentists should never reveal confidential information to the police unless they have the relevant paperwork etc. There is however, excellent discussion of the consequences of breaking confidentiality and a good level of detail (e.g. Public health act 1988).

Conclusion: An excellent conclusion that not only summarises the main arguments from both sides but also builds upon these to offer a solution as to when to break confidentiality by treating it on a case-by-case basis.

Language: There is only one spelling mistake ("*necessry*") *and a*lthough the somewhat general phrases stop this from being a perfect A5, it is still an excellent essay that displays good insight. Spend time making sure that you write exactly what you mean, instead of being loose with your words and conveying an incorrect message.

Score: A4.5

For 12 more BMAT example essays check the *Ultimate BMAT Guide*– flick to the back to get a free copy.

Summary

| **Intro** | • Does it explain or just repeat?
• Does it set up the main body?
• Does it get to the point? |

| **Main Body** | • Are enough points being made? *[Breadth]*
• Are the points explained sufficiently? *[Depth]*
• Does the argument make sense? *[Strength]* |

| **Conclusion** | • Does it follow naturally from the main body?
• Does it consider both sides of the argument?
• Does it answer the original question? |

General Advice

✓ Always answer the question clearly – this is the key thing examiner look for in an essay.

✓ Analyse each argument made, justifying or dismissing with logical reasoning.

✓ Keep an eye on the time/space available – an incomplete essay may be taken as a sign of a candidate with poor organisational skills.

✓ Use pre-existing knowledge when possible – examples and real world data can be a great way to strengthen an argument- but don't make up statistics!

✓ Present ideas in a neat, logical fashion (easier for an examiner to absorb).

✓ Complete some practice papers in advance, in order to best establish your personal approach to the paper (particularly timings, how you plan etc.).

✗ Attempt to answer a question that you don't fully understand, or ignore part of a question.

✗ Rush or attempt to use too many arguments – it is much better to have fewer, more substantial points.

✗ Attempt to be too clever, or present false knowledge to support an argument – a tutor may call out incorrect facts etc.

✗ Panic if you don't know the answer the examiner wants – there is no right answer, the essay is not a test of knowledge but a chance to display reasoning skill.

✗ Leave an essay unfinished – if time/space is short, wrap up the essay early in order to provide a conclusive response to the question.

BMAT ANSWERS

Q	A	Q	A	Q	A	Q	A	Q	A	Q	A
1	A	16	B	31	B	46	C	61	A	76	C
2	C	17	E	32	C	47	D	62	C	77	C
3	A	18	E	33	D	48	B	63	F	78	E
4	A	19	D	34	A	49	A	64	A	79	A
5	C	20	E	35	D	50	D	65	D	80	C
6	D	21	E	36	B	51	D	66	E	81	E
7	D	22	A	37	D	52	A	67	G	82	E
8	A	23	C	38	D	53	D	68	C	83	C
9	A	24	D	39	C	54	E	69	D	84	E
10	B	25	E	40	C	55	A	70	E	85	E
11	D	26	C	41	A	56	A	71	D		
12	C	27	B	42	F	57	D	72	C		
13	D	28	E	43	F	58	C	73	F		
14	A	29	D	44	A	59	B	74	B		
15	D	30	C	45	C	60	F	75	C		

Worked Answers

Question 1: A
Whilst **B**, **C** and **D** may be true, they are not completely stated, **A** is clearly stated and so is the correct answer.

Question 2: C
The main argument of the first paragraph is to propose the point that it is more society that controls gender behaviour not genetics. **A** and **D** do not indicate either as they only allude to the end result of gender behaviour and so are incorrect. Hormonal effects are not mentioned in the first paragraph and so **B** is incorrect. **C** would undermine the argument that society *predominately* controls gender, and so is correct.

Question 3: A
B, C and D are not stated and so are incorrect. A is directly stated and so is correct.

Question 4: A
B and **D** are contraindicated by the statement and so are incorrect. **C** could be true but implies children always like the same thing as their same-gendered parent irrelevant of how they are treated as a child, which is contrary to the statement and so is not correct. **A** is correct as is the overall message.

Question 5: C
D may help prevent problems with sexual identity but does not prevent stereotyping and so is incorrect. **A** is not stated, and **B** is implied but not stated and so are incorrect. **C** is the end message of how to prevent gender stereotyping and so is correct.

Question 6: D
A, B and C may be true but are not mentioned in the statement and so are incorrect. The statement implies that children born with different external organs to those that their sex chromosomes would match may find it difficult to accept this difference and be uncomfortable.

Question 7: D
The text states that 'Those who regularly took 30-minute naps were more than twice as likely to remember simple words such as those of new toys.' Which means those who napped were twice as likely to remember teddy's name than the 5% who did not, 5% x 2 = 10%, which would be twice as likely, ruling out **A** and **B**. But being 'more than twice' the only possible answer is **D**.

Question 8: A
The answer is to work out 10% (the percentage of napping toddlers more likely to suffer night disturbances) of 75% (the percentage of toddlers who regularly nap). Hence 10 % of 75% is 7.5%.

Question 9: A
B, **C** and **D** may be true but there is nothing in the text to support them. **A** is suggested, as the passage states 'non-napping counterparts, who also had higher incidences of memory impairment, behavioural problems and learning difficulties'. If the impaired memory were the cause, as opposed to the result, of irregular sleeping then it would offer an alternative reason why those who nap less remember less.

Question 10: B
A and **C** are possible implications but not stated and so are incorrect. It is said that parents cite napping having 'the benefits of their child having a regular routine' so hence **B** is more correct than **D** as it refers to the benefit to the toddlers' rather than the parents.

Question 11: D
B, if true would counteract the conclusion, as it would imply that, the study is skewed. The same is true of **C**, which if true would imply unreliable results as the toddler sample are all the same age within a year, but not within a few weeks. **A**, if true, would not provide any additional support to the conclusion and so is incorrect. **D** if true would provide the most support for the conclusion as it proposes using groups with a higher incidence of napping in comparison to those with a lower incidence.

Question 12: C
Although it can be argued that **A**, **B**, **D** and **E** are true they are not the best answer to demonstrate a flaw in Tom's father's argument. **C** is the best because it accounts for other factors determining success for the Geography A-level exam such as aptitude for the subject.

Question 13: D
A is never stated and is incorrect. **B** and **C** are referred to being 'many people's' beliefs, and are cited as others' opinions not an argument supported by evidence in the passage, and so are not valid conclusions. It is implied that the NHS may have to reduce its services in the future, some of which could be fertility treatments hence **D** is the most correct answer.

Question 14: A
C does not severely affect the strength of the argument, as it is only relevant to the length of the time taken for the effects of the argument to come into place.
D is incorrect, as people breaking speed limits already would not negate the argument that speed limits should be removed, but could even be seen as supporting it. These people may count as the 'dangerous drivers' who would be ultimately weeded out of the population.

B may affect part of the argument's logic (as it undermines the idea that dangerous drivers are born to dangerous drivers), but the final conclusion that dangerous drivers will end up killing only themselves still stands, and so the ultimate population of only safe drivers may be obtained. The fact that one dead dangerous driver could have produced a safe one does not necessarily challenge the main point of this argument.
A if true would most weaken the argument as it states that fast driver is more likely to harm others and not the driver itself, which would negate the whole argument.

Question 15: D

The easiest thing to do is draw the relative positions. We know Harrington is north of Westside and Pilbury. We know that Twotown is between Pilbury and Westside. Crewville is south of Twotown, Westside and Harrington but we do not know but its location relative to Pilbury.

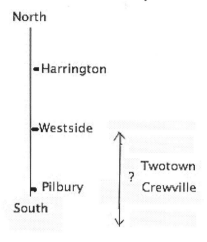

Question 16: B

By making a grid and filling in the relevant information the days Dr James works can be deduced:

	Sunday	Monday	Tuesday	Wednesday	Thursday	Friday	Saturday
Dr Evans	X	√	X	X	√	√	√
Dr James	X	√	√	√	√	X	√
Dr Luca	X	X	√	√	X	√	√

- No one works Sunday.
- All work Saturday.
- Dr Evans works Mondays and Fridays.
- Dr Luca cannot work Monday or Thursday.
- So, Dr James works Monday.
- And, Dr Evans and Dr James must work Thursday.
- Dr Evans cannot work 4 days consecutively so he cannot work Wednesday.
- Which means Dr James and Luca must work Wednesday.
- (mentioned earlier in the question) Dr Evans only works 4 days, so cannot work Tuesday.
- Which means Dr James and Luca work Tuesday.
- Dr James cannot work 5 days consecutively so cannot work Friday.
- Which means Dr Luca must work Friday.

Question 17: E

Working algebraically, using the call out rate as C, and rate per mile as M.

So,   C + 4m = 11

C + 5m = 13

Hence; (C + 5m) – (C + 4m) = £13 - £11

M = £2

Substituting this back into C + 4m = 11

C + (4 x 2) = 11

Hence, C = £3

Thus a ride of 9 mile will cost £3 + (9 x £2) = £21.

Question 18: E

Use the information to create a Venn diagram.

We don't know the exact position of both Trolls and Elves, so **A** and **D** are true. Goblins are mythical but not magical, so **C** is true. Gnomes are neither so **B** is true. But **E** is not true.

Question 19: D

The best method may be work backwards from 7pm. The packing (15 minutes) of all 100 tiles must have started by 6:45pm, hence the cooling (20 minutes) of the last 50 tiles started by 6:25pm, and the heating (45 minutes) by 5:40pm. The first 50 heating (45 minutes) must have started by 4:35pm, and cooling (20 minutes) by 5:20pm. The decoration (50 minutes) of the second 50 can occur anytime during 4:35pm-5:40pm as this is when the first 50 are heating and cooling in the kiln, and so does not add time. The first 50 take 50 minutes to decorate and so must be started by 3:45pm.

Question 20: E

Speed = distance/time. Hence for the faster, pain impulse the speed is 1m/ 0.001 seconds. Hence the speed of the pain impulse is 1000 metres per second. The normal touch impulse is half this speed and so is 500 metres per second.

Question 21: E

Using the months of the year, Melissa could be born in March or May, Jack in June or July and Alina in April or August. With the information that Melissa and Jack's birthdays are 3 months apart the only possible combination is March and June. Hence Alina must be born in August, which means it is another 7 months until Melissa's birthday in March.

Question 22: A

PC Bryan cannot work with PC Adams because they have already worked together for 7 days in a row, so **C** is incorrect. **B** is incorrect because if PC Dirk worked with PC Bryan that would leave PC Adams with PC Carter who does not want to work with him. PC Carter can work with PC Bryan.

Question 23: C

Paying for my next 5 appointments will cost £50 per appointment before accounting for the 10% reduction, hence the cost counting the deduction is £45 per appointment. So the total for 4 appointments = 5 x £45 = £225 for the hair. Then add £15 for the first manicure and £10 x 2 for the subsequent manicures using the same bottle of polish bringing an overall total of £260.

Question 24: D

Elena is married to Alex or David, but we are told that Bertha is married to David and so Alex must be married to Elena. Hence David, Bertha, Elena and Alex are the four adults. Bertha and David's child is Gemma. So Charlie and Frankie must be Alex and Elena's two children. Leaving only options **A** or **D** as possibilities. Only Frankie and Gemma are girls so Charlie must be a boy.

Question 25: E

A. Incorrect. UCL study found eating more portions of fruit and vegetables was beneficial.
B. Incorrect. This is a possible reason but has yet to be fully investigated.
C. Incorrect. Fruit and vegetables are more protective against cardiovascular disease, and were shown to have little effect on cancer rates.
D. Incorrect. Inconclusive – people who ate more vegetables generally had a lower mortality but unknown if this is due to eating more vegetables or other associated factors.
E. Correct. Although this has previously been the case, this study did not find so. 'they recorded no additional decline for people who ate over 5 portions'.
F. Incorrect. The 5% decline per portion was only up to 5 portions and no additional reduction in mortality for 7 than 5 portions.
G. Incorrect. Study only looks at cancers in general and states need to look into specific cancers.

Question 26: C
Deaths in meta-analysis = 56423/800000 = 0.07 or 7%
1% lower in UCL study so 6%
6% of 65,000 = 65000 x 0.06 = 3,900

Question 27: B
A. Eating more fruit and vegetables doesn't particularly lower overall risk but need research into specific cancer risk.
B. The UCL research alone found that increasing the number of fruit and vegetable portions had a beneficial effect, even though this wasn't the overall conclusion when combined with results from the meta-analysis.
C. The results were not exactly the same but showed similar overall trends.
D. Although this may be true, there is no mention of this in the passage.
E. Fruit and vegetables are protective against cardiovascular disease, but not exclusively. They also reduce the rates of death from all causes.
F. The UCL study is in England only and the meta-analysis a combination of studies from around the world.
G. Suggested by the UCL research, but not the meta-analysis, so not an overall conclusion of the article.

Question 28: E
Remember that you don't need to calculate exact values for question 28 – 30. Thus, you should round numbers frequently to make this more manageable. Work out percentage of beer and wine consumption and then the actual value using the total alcohol consumption figure:

Belarus: 17.3 + 5.2 = 22.5%;
0.225 x 17.5 = 3.94

Lithuania: Missing figure 100 – 7.8 – 34.1 – 11.6 = 46.5
46.5 + 7.8 = 54.3%
0.543 x 15.4 = 8.36

France: 18.8 + 56.4 = 75.2%
0.752 x 12.2 = 9.17

Ireland: 48.1 + 26.1 = 74.2
0.742 x 11.9 = 8.83

Andorra: missing figure 100 – 34.6 – 20.1 = 45.3
34.6 + 45.3 = 79.9%
0.799 x 13.8 = 11.0

Question 29: D

Russia:
2010 – Total = 11.5+3.6 = 15.1. Spirits = 0.51 x 15.1 = 7.7
2020 – Total = 14.5. Spirits = 0.51 x 14.5 = 7.4
Difference = 0.3 L

Belarus:
2010 – Total = 14.4 + 3.2 = 17.6. Spirits = 0.466 x 17.6 = 8.2
2020 – Total = 17.1. Spirits = 0.466 x 17.1 = 8.0
Difference = 0.2 L

Lithuania:
2010 – Total = 15.4. Spirits = 0.341 x 15.4 = 5.3
2020 – Total = 16.2. Spirits = 0.341 x 16.2 = 5.5
Difference = 0.2 L

Grenada:
2010 – Total = 12.5. Spirits % = 100 – 29.3 – 4.3 – 0.2 = 66.2%. Spirits = 0.662 x 12.5 = 8.3

2020 – Total = 10.4. Spirits = 0.662 x 10.4 = 6.8
Difference = 1.5 L

Ireland:
2010 – Total = 11.9. Spirits = 0.187 x 11.9 = 2.2
2020 – Total = 10.9. Spirits = 0.187 x 10.9 = 2
Difference = 0.2 L

Question 30: C
Work out 4.9 as a percentage of total beer consumption in Czech Republic and search other rows for similar percentage.

4.9/13 = 0.38, approx. 38% which is very similar to percentage consumption in Russia (37.6).

Question 31: B
We can add up the total incidence of the 6 cancers in men, which is 94,000. Then we can add up the total incidence in women, which is 101,000. As a percentage of 10 million, this is 0.94% of men and 1.01% of women. Therefore the difference is 0.07%.

Question 32: C
Given there are 1.15 times as many men as women, the incidence of each cancer amongst men needs to be greater than 1.15 times the incidence amongst women in order for a man to be more likely to develop it. The incidence is at least 1.15 higher in men for 3 cancers (prostate, lung and bladder).

Question 33: D
If 10% of cancer patients are in Sydney, there are 10,300 prostate/bladder/breast cancer patients and 9,200 lung/bowel/uterus cancer patients in Sydney. Hence the total number of hospital visits is 10,300 + 18,400, which is 28,700.

Question 34: A
The proportion of men with bladder cancer is 2/3 and women 1/3.

Question 35: D
First we work out the size of each standard drink. 50 standard drinks of vodka is equivalent to 1250ml, so one drink is 25ml or 0.025 litres. 11.4 standard drinks of beer is 10 pints of 5700ml, so one standard drink is 500ml or 0.5 litres. 3 standard drinks of cocktail is 750ml so one is 250ml or 0.25 litres. 3.75 standard drinks of wine is 750ml, so one is 200ml or 0.2 litres.
We can then work out the number of units in each drink. Vodka has 0.025 x 40 = 1 unit, Beer has 0.5 x 3 = 1.5 units, Cocktail has 0.25 x 8 = 2 units and Wine has 0.2 x 12.5 = 2.5 units. Since the drink with the most units is wine, the answer is D.

Question 36: B
We found in the last question that vodka has 1 unit, beer has 1.5, cocktail has 2 and wine has 2.5. Hence in the week, Hannah drinks 23.5 units and Mark drinks 29 units. Hence Hannah exceeds the recommended amount by 9.5 units and Mark by 9 units.

Question 37: D
We found that vodka has 1 unit, beer has 1.5, cocktail has 2 and wine has 2.5. Hence it is possible to make 5 combinations of drinks that are 4 units: 4 vodkas, 2 cocktails, 2 vodkas and a cocktail, 1 vodka and 2 beers, or a wine and a beer.

Question 38: D
The total number of males in Greentown is 12,890. Adding up the rest of the age categories, we can see that 10,140 of these are in the older age categories. Hence there are 2750 males under 20.

Question 39: C

Given that in the first question we found the number of males under 20 is 2,750, we can then add up the totals in the age categories (apart from 40-59) in order to find that 15,000 of the residents of Greentown are in other age categories. Hence 9,320 of the population are aged 40-59. We know that 4,130 of these are male, therefore 5,190 must be female.

Question 40: C

The age group with the highest ratio of males:females is 20-39, with approximately 1.9 males per females (approximately 3800:2000). As a ratio of females to males, this is 1:1.9.

Question 41: A

DNA consists of 4 bases: adenine, guanine, thymine and cysteine. The sugar backbone consists of deoxyribose, hence the name DNA. DNA is found in the cytoplasm of prokaryotes.

Question 42: F

Mitochondria are responsible for energy production by ATP synthesis. Animal cells do not have a cell wall, only a cell membrane. The endoplasmic reticulum is important in protein synthesis, as this is where the proteins are assembled.

Question 43: F

If you aren't studying A-level biology, this question may stretch you. However, it is possible to reach an answer by process of elimination. Mitochondria are the 'powerhouse' of the cell in aerobic respiration, responsible for cell energy production rather than DNA replication or protein synthesis. As energy producers they are required in muscle cells in large numbers, and in sperm cells to drive the tail responsible for movement. They are enveloped by a double membrane, possibly because they started out as independent prokaryotes engulfed by eukaryotic cells.

Question 44: A

The majority of bacteria are commensals and don't lead to disease.

Question 45: C

Bacteria carry genetic information on plasmids and not in nuclei like animal cells. They don't need meiosis for replication, as they do not require gametes. Bacterial genomes consist of DNA, just like animal cells.

Question 46: C

Active transport requires a transport protein and ATP, as work is being done against an electrochemical gradient. Unlike diffusion, the relative concentrations of the materials being transported aren't important.

Question 47: D

Meiosis produces haploid gametes. This allows for fusion of 2 gametes to reach a full diploid set of chromosomes again in the zygote.

Question 48: B

Mendelian inheritance separates traits into dominant or recessive. It applies to all sexually reproducing organisms. Don't get confused by statement C – the offspring of 2 heterozygotes has a 25% chance of expressing a recessive trait, but it will be homozygous recessive.

Question 49: A

Hormones are released into the bloodstream and act on receptors in different organs in order to cause relatively slow changes to the body's physiology. Hormones frequently interact with the nervous system, e.g. Adrenaline and Insulin, however, they don't directly cause muscles to contract. Almost all hormones are synthesised.

Question 50: D

Neuronal signalling can happen via direct electrical stimulation of nerves or via chemical stimulation of synapses which produces a current that travels along the nerves. Electrical synapses are very rare in mammals, the majority of mammalian synapses are chemical.

Question 51: D

Remember that pH changes cause changes in electrical charge on proteins (= polypeptides) that could interfere with protein – protein interactions. Whilst the other statements are all correct to a certain extent, they are the downstream effects of what would happen if enzymes (which are also proteins) didn't work.

Question 52: A

The bacterial cell wall is made up of cellulose and protects the bacterium from the external environment, in particular from osmotic stresses, and is important in most bacteria.

Question 53: D

Different isotopes are differentiated by the number of neutrons in the core. This gives them different molecular weights and different chemical properties with regards to stability. The number of protons defines each element, and the number of electrons its charge.

Question 54: E

A displacement reaction occurs when a more reactive element displaces a less reactive element in its compound. Reaction 4 will not happen as lead is less reactive than sodium

Question 55: A

There needs to be 3Ca, 12H, 14O and 2P on each side. Only option A satisfies this.

Question 56: A

To balance the equation there needs to be 9Ag, 9N, $9O_3$, 9K, 3P on each side. Only option A satisfies this.

Question 57: D

A more reactive halogen can displace a less reactive halogen. Thus, chlorine can displace bromine and iodine from an aqueous solution of its salts, and fluorine can replace chlorine. The trend is the opposite for alkali metals, where reactivity increases down the group as electrons are further from the core and easier to lose.

Question 58: C

$2Mg + O_2 = 2MgO$

so 2 x 24 = 48 and 2 x (24 + 16) = 80

so 48 g of magnesium produces 80g of magnesium oxide

so 1g of magnesium produces 1g x 80g/48g = 1.666g oxide

so 75g x 1.666 = 125g

Question 59: B

$H_2 + 2OH^- \rightarrow 2H_2O + e^-$

Thus, the hydrogen loses electrons i.e. is oxidised.

Question 60: F

Ammonia is 1 nitrogen and 3 hydrogen atoms bonded covalently. N = 14g and H = 1g per mole, so percentage of N in NH_3 = 14g/17g = 82%. It can be produced from N_2 through fixation or the industrial Haber process for use in fertiliser, and may break down to its components.

Question 61: A

Milk is weakly acidic, pH 6.5-7.0, and contains fat. This is broken down by lipase to form fatty acids - turning the solution slightly more acidic.

Question 62: C

Glucose loses four hydrogen atoms; one definition of an oxidation reaction is a reaction in which there is loss of hydrogen.

Question 63: F

That the amplitude of a wave determines its mass is false. Waves are not objects and do not have mass.

Question 64: A

We know that displacement s = 30 m, initial speed u = 0 ms^{-1}, acceleration a = 5.4 ms^{-2}, final speed v = ?, time t = ?

And that $v^2 = u^2 + 2as$

$v^2 = 0 + 2 \times 5.4 \times 30$

$v^2 = 324$ so v = 18 ms^{-1}

and s = ut + 1/2 at^2 so 30 = 1/2 $\times$ 5.4 $\times$ t^2

t^2 = 30/2.7 so t = 3.3 s

Question 65: D

The wavelength is given by: velocity v = λf and frequency f = 1/T so v = λ/T giving wavelength λ = vT

The period T = 49 s/7 so λ = 5 ms^{-1} $\times$ 7 s = 35 m

Question 66: E

This is a straightforward question as you only have to put the numbers into the equation (made harder by the numbers being hard to work with).

$Power = \frac{Force \times Distance}{Time} = \frac{375\ N \times 1.3\ m}{5\ s}$

$= 75 \times 1.3 = 97.5\ W$

Question 67: G

v = u + at

v = 0 + 5.6 $\times$ 8 = 44.8 ms^{-1}

And $s = ut + \frac{at^2}{2} = 0 + 5.6 \times \frac{8^2}{2} = 179.2$

Question 68: C

The sky diver leaves the plane and will accelerate until the air resistance equals their weight – this is their terminal velocity. The sky diver will accelerate under the force of gravity. If the air resistance force exceeded the force of gravity the sky diver would accelerate away from the ground, and if it was less than the force of gravity they would continue to accelerate toward the ground.

Question 69: D

s = 20 m, u = 0 ms^{-1}, a = 10 ms^{-2}

and $v^2 = u^2 + 2as$

$v^2 = 0 + 2 \times 10 \times 20$

$v^2 = 400$;   v = 20 ms^{-1}

Momentum = Mass $\times$ velocity = 20 $\times$ 0.1 = 2 kgms^{-1}

Question 70: E

Electromagnetic waves have varying wavelengths and frequencies and their energy is proportional to their frequency.

Question 71: D

The total resistance = R + r = 0.8 + 1 = 1.8 Ω

and $I = \frac{e.m.f}{total\ resistance} = \frac{36}{1.8} = 20\ A$

Question 72: C

Use Newton's second law and remember to work in SI units:

So $Force = mass \times accelaration = mass \times \frac{\Delta velocity}{time}$

$= 20 \times 10^{-3} \times \frac{100 - 0}{10 \times 10^{-3}}$

$= 200\ N$

Question 73: F

In this case, the work being done is moving the bag 0.7 m

i.e. $Work\ Done\ =\ Bag's\ Weight\ x\ Distance\ =\ 50\ x\ 10\ x\ 0.7 = 350\ N$

$Power = \frac{Work}{Time} = \frac{350}{3} = 116.7\ W$

$= 117\ W$ to 3 significant figures

Question 74: B

Each three-block combination is mutually exclusive to any other combination, so the probabilities are added. Each block pick is independent of all other picks, so the probabilities can be multiplied. For this scenario there are three possible combinations:

P(2 red blocks and 1 yellow block) = P(red then red then yellow) + P(red then yellow then red) + P(yellow then red then red) =

$(\frac{12}{20}\ x\frac{11}{19}\ x\frac{8}{18}) + (\frac{12}{20}\ x\frac{8}{19}\ x\frac{11}{18}) + (\frac{8}{20}\ x\frac{12}{19}\ x\frac{11}{18}) =$

$\frac{3\ x\ 12\ x\ 11\ x\ 8}{20\ x\ 19\ x\ 18} = \frac{44}{95}$

Question 75: C

Multiply through by 15: $3(3x + 5) + 5(2x - 2) = 18\ x\ 15$

Thus: $9x\ +\ 15 + 10x\ -\ 10\ =\ 270$

$9x\ +\ 10x\ =\ 270\ -\ 15 + 10$

$19x\ =\ 265$

$x\ =\ 13.95$

Question 76: C

This is a rare case where you need to factorise a complex polynomial:

(3x   )(x   ) = 0, possible pairs: 2 x 10, 10 x 2, 4 x 5, 5 x 4

(3x - 4)(x + 5) = 0

3x - 4 = 0, so x = $\frac{4}{3}$

x + 5 = 0, so x = -5

Question 77: C

$\frac{5(x-4)}{(x+2)(x-4)} + \frac{3\ (x+2)}{(x+2)(x-4)}$

$= \frac{5x-20+\ 3x+6}{(x+2)(x-4)}$

$= \frac{8x-14}{(x+2)(x-4)}$

Question 78: E

p α $\sqrt[3]{q}$, so p = k $\sqrt[3]{q}$

p = 12 when q = 27 gives 12 = k $\sqrt[3]{27}$, so 12 = 3k and k = 4

so p = 4 $\sqrt[3]{q}$. Now p = 24:

24 = 4$\sqrt[3]{q}$, so 6 = $\sqrt[3]{q}$ and q = $6^3 = 216$

Question 79: A

8 x 9 = 72; 8 = (4 x 2) = 2 x 2 x 2; 9 = 3 x 3

(2 x 2 x 2 x 3 x 3)^2 = 2 x 2 x 2 x 2 x 2 x 2 x 3 x 3 x 3 x 3 = 2^6 x 3^4

Question 80: C

Note that 1.151 x 2 = 2.302.

Thus: $\frac{2 \; x \; 10^5 + 2 \; x \; 10^2}{10^{10}} = 2 \; x \; 10^{-5} + 2 \; x \; 10^{-8}$

= 0.00002 + 0.00000002 = 0.00002002

Question 81: E

y^2 + ay + b

= (y +2)^2 - 5 = y^2 + 4y + 4 - 5

= y^2 + 4y + 4 - 5 = y^2 + 4y - 1

So a = 4 and y = -1

Question 82: E

Take $5(m + 4n)$ as a common factor to give: $\frac{4(m+4n)}{5(m+4n)} + \frac{5(m-2n)}{5(m+4n)}$

Simplify to give: $\frac{4m+16n+5m-10n}{5(m+4n)} = \frac{9m+6n}{5(m+4n)} = \frac{3(3m+2n)}{5(m+4n)}$

Question 83: C

$A \; \alpha \; \frac{1}{\sqrt{B}}$. Thus, $= \frac{k}{\sqrt{B}}$.

Substitute the values in to give: $4 = \frac{k}{\sqrt{25}}$.

Thus, $k = 20$. Therefore, $A = \frac{20}{\sqrt{B}}$.

When B = 16, $A = \frac{20}{\sqrt{16}} = \frac{20}{4} = 5$

Question 84: E

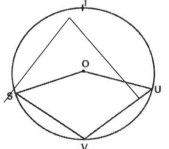

Angles SVU and STU are opposites and add up to 180°, so STU = 91°

The angle of the centre of a circle is twice the angle at the circumference so SOU
= 2 x 91° = 182°

Question 85: E

The surface area of an open cylinder A = 2πrh. Cylinder B is an enlargement of

A, so the increases in radius (r) and height (h) will be proportional: $\frac{r_A}{r_B} = \frac{h_A}{h_B}$. Let us call the proportion

coefficient n, where n $= \frac{r_A}{r_B} = \frac{h_A}{h_B}$.

So $\frac{Area \; A}{Area \; B} = \frac{2\pi r_A h_A}{2\pi r_B h_B} = n \; x \; n = n^2$. $\frac{Area \; A}{Area \; B} = \frac{32\pi}{8\pi} = 4$, so n = 2.

The proportion coefficient n = 2 also applies to their volumes, where the third dimension (radius, i.e. the r^2 in

V = πr^{2}h) is equally subject to this constant of proportionality. The cylinder's volumes are related by n^3 = 8.

If the smaller cylinder has volume 2π cm^3, then the larger will have volume 2π x n^3 = 2π x 8 = 16π cm^3.

DENTAL SCHOOL INTERVIEWS

Getting a dental school interview is a great achievement – well done if you get one. They can be a daunting prospect, but with the right preparation, there is every reason to be confident that you can present yourself in the best possible way.

This section is aimed at giving you a comprehensive walkthrough of the entire interview process- from the very basics of your initial preparation to the moment you walk out of the interview room.

What is a Dental Interview?

A dental interview is a formal discussion between interviewer and interviewee. This is normally the *final* step in the dental school application process. Traditionally, interviews last around 20 – 30 minutes and take place from mid-October to April every year.

Why is there an Interview?

Dentistry is competitive- the vast majority of applicants will already have outstanding grades at GCSE, fantastic predicted exam results, and a solid personal statement. As you can imagine, this makes many applicants look very similar on paper!

The interview process is designed to identify students that would be best suited for studying dentistry at the university you've applied to. Thus, dental school interviews assess multiple qualities e.g. your motivation to study dentistry, communication skills, team working and leadership skills etc.

Whilst the interview is not testing your knowledge and skills about a specific subject, it helps to have a good knowledge base. The interview process will assess candidates on their ability to learn. The universities are looking for candidates that have the ability and mindset to learn, i.e. can the candidate be taught. Universities recognise that the interviews can be stressful and may utilise this to assess candidates' reactions to stress; to see if they can cope with the pressure. It is a process where the candidates are judged on their quick thinking, logical approach, and ability to formulate a comprehensive, coherent, structured response. Your interviewer will guide you through the process. The interview is not about tricking you. It's about testing your abilities to harp on your existing knowledge and use it to come up with a logical response. The interview is also about assessing your abilities to see if you can come up with plausible solutions, even if you don't know the answer.

Who gets an Interview?

Interviews are usually the last step in the application process. Each university has its own unique selection process and will value some parts of the admissions process more than others. However, most will only interview around 30-60% of applicants – the rest are rejected without interview (although there are some notable exceptions).

Whether or not you get an interview invite will depend on how well you satisfy each university's entrance criteria. This is normally a combination of predicted grades, your personal statement, school reference, UKCAT/BMAT scores, etc.

It is important to research your universities early on to get a better idea of their particular preferences. For example, Birmingham will generally only interview students who attain 8 or more A*s at GCSE; Kings will generally only students with very high UKCAT scores.

Who are the Interviewers?

Depending on the university, dental school interviews can be conducted by a host of different people:

- Dean of the Dental School
- Senior Dentists and Lecturers
- Nurses
- Other Allied Health Professionals
- Current Dental Students

When is the Interview?

This will vary depending on the university you've applied to. Generally, most universities will start interviewing in late October/early November and continue to do so until January. Some universities will carry on interviewing until early May.

Most universities will give you at least 2 weeks' notice prior to your interview. However, it's highly advisable to **begin preparing for your interview before you're officially invited**.

Can I Change the Interview Date?

It's generally not a good idea to change your interview date unless you absolutely have to and have a strong reason to do so, e.g. Family Bereavement. Rescheduling your interview for your friend's birthday or a sports match is unlikely to go down well!

If you do need to reschedule, give the dental school as much notice as possible and offer some alternative dates. Be aware that you may be putting yourself at a disadvantage by doing of this, as your interview will likely be delayed.

Where is the Interview?

The interview will almost always take place at the dental school you've applied to. Logistically, it's worth booking an overnight hotel if you're travelling from far away.

Very rarely, interviews can be held via Skype at an exam centre- this normally only applies to international students or for UK students in extreme circumstances.

What Should I Wear?

When it comes to interviews, it's best to dress sharply and smartly.

Unless you're told otherwise by the dental school, this normally means a full suit for men and either a full suit or smart shirt + skirt for women.

Things to do:
- Carry an extra pair of contact lenses or glasses if appropriate
- Turn your phone off completely – you don't want any distractions
- Polish your shoes
- Avoid bringing a large bag
- Avoid excessive or intricate make-up, nails or jewellery – clean and professional is the aim

Body Language

First impressions last; body language contributes to a significant part of this. However, don't make the mistake of obsessing over body language at the expense of the quality answers you give.

Most people will only need to make some minor adjustments to remove some "bad habits", so don't worry about body language till fairly late in your preparation.

Once you're confident that you know the relevant material and have good answers prepared for common questions, allow your body language to show that you have what it takes to be a dentist by conveying maturity and confidence.

Posture
➢ When walking into the room, walk in with your head held high and back straight.
➢ When sitting down, look alert and sit up straight.
➢ Avoid crossing your arms – this can appear to be defensive.
➢ Don't slouch- instead, lean forward slightly to show that you're engaged with the interview.
➢ If there is a table, then ensure you sit around four to six inches away.
  ❖ Too close and you'll appear like you're invading the interviewers' space
  ❖ Too far and you'll appear too casual

Eyes
➢ Good eye contact is a sign of confidence and good communication skills.
➢ Look at the interviewer when they are speaking to you and when you are speaking.
➢ If there are multiple interviewers, look at the interviewer who is speaking to you or asked you the question. However, make sure you do look around at the other interviewers to acknowledge them.

Hands
➢ At the start, offer a handshake or accept if offered: make sure you don't have sweaty or cold hands.
➢ A firm handshake is generally preferable to a limp one.
➢ During the interview, keep your hands still unless you are using them to illustrate your point.
➢ Avoid excessive hand movements – your hands should go no higher than your neck.
➢ If you fidget when you're nervous, hold your hands firmly together in your lap to stop this from happening.

Top Tip! Make sure you practise speaking articulately about the topics that interest you. Join your school's debating society (or set one up!), discuss it with friends & family, challenge your biology teacher regarding a point you don't quite understand or would like to know more about!

Preparing for Interviews

How do I Prepare?

Many applicants will jump straight to the questions part of this book and attempt to learn the "good answers" by rote. Whilst this is psychologically comforting, it is of minimal value in actually preparing you for the interview. Use this book as guidance. Develop the skills to construct logical and structured answers so that you are familiar with the format. Cramming for your interview by rote learning the answers provided in this book will make your responses clichéd.

You are almost guaranteed to be asked certain questions at your interview, e.g. "Why Dentistry?". So it's well worth preparing answers for these frequently asked questions. However, it's critical that you don't simply recite pre-prepared answers as this will appear unnatural and, therefore, rehearsed.

Step One: Do your Research
Start off by finding out exactly what is required of you for the interview, e.g. interview type, who your interviewers will be, commonly asked questions, etc.

Step Two: Acquire the pre-requisite Knowledge
Next, focus on learning the basics of interview technique and understanding the core dentistry topics that you'll be expected to know about. Once you're happy with this, go through the questions + answers in this book – don't try to rote learn answers. Instead, try to understand what makes each answer good/bad and then use this to come up with your own unique answer. Don't be afraid of using other resources, e.g. YouTube videos on dental ethics, etc.

When you've got this down, practice answering questions in front of a mirror and consider recording yourself to iron out any body language issues.

Step Three: Practice with People
It's absolutely paramount that you practice with a real person before your first interview. Whilst you might be able to provide fantastic answers to common questions with no one observing you, this may not be the case with the added pressure of a mock interviewer.

Practice interviews are best with someone you do not know very well - even easy questions may be harder to articulate out loud and on the spot to a stranger. MMIs (Multiple Mini Interview), in particular, are worth practising beforehand, so you can work on using the time available as efficiently as possible. During your practice, try to eliminate hesitant words like "Errrr…." and "Ummm…" as these will make you appear less confident. Ask for feedback on the speed, volume, and tone of your voice.

Many schools will be able to arrange a mock interview for you. If you're struggling, you can book private mock interviews at www.uniadmissions.co.uk.

> *Top Tip!* Practise structuring arguments in your head so that you can present them in a logical and easy-to-follow way. This will show that you have previously thought carefully about the topic in hand.

Types of Interviews

The first thing to understand is that there are several different formats of an interview. They can broadly be categorised into:

- Panel-Based Interview:
  - Multiple interviewers
  - Normally twenty – forty minutes
  - May have two long panel interviews
- Multiple Mini Interview (MMI):
  - There are normally five to twelve stations
  - Each station is five to ten minutes long and will focus on different topic
  - There is normally an interviewer at each station
  - You physically move from one station to another
  - You will often be provided with a scenario before entering
  - You will have about 1 minute to read the scenario background before beginning

Different dental schools approach interviews in different ways. Some may concentrate more on dissecting your personal statement, exploring your motivations and dental work experience. Others may try to test you primarily on dental ethics whilst the BMAT dental schools may question you on your BMAT essay. Hence, it's extremely important that you know what type of interview you will be going through as that determines how you prepare.

Dental schools will let you know the exact interview format in advance; some will also tell you who your interviewers will be. It can be useful to **look at your interviewers' teaching backgrounds and published work** as this can potentially shed some light on the topics they might choose to discuss during the interview. However, there is absolutely no need to know the intricacies of their research work so don't get bogged down in it. It can be useful to know their views on their areas of interest so that you are prepared and can offer both sides of the argument in a balanced manner.

Interviews tend to open with easier and more general questions and become more detailed and complicated as you are pushed to explore topics in greater depth. Remember, if the questions are getting harder, you are probably doing well!

The whole point of interviews is to identify individuals who will fit in at that institution, so your grades, interest, hobbies, and experiences are all important and you may be asked on any of these. Also, be prepared to discuss your personal statement, current affairs, and your BMAT essay if your university requires it.

Generally, MMIs tend to be highly scripted with each interviewer having a certain number of questions that they need to ask in the time limit. Contrastingly, traditional interviews are more free flowing – there are usually lots of follow-up questions based on your previous responses as there is more time to explore more complex issues. An example of that is a discussion that starts with a simple question like: *What did you enjoy in your work experience?*

Although the format of dental school interviews can vary considerably, the qualities which interviewers are looking for in applicants remain consistent. It's well worth your time to understand these qualities and exhibit them as much as possible when answering questions.

Top Tip! Read! If you're genuinely interested in dentistry then you should love reading around it. Read something accessible to you (i.e. **NOT** Shillingburg's Fundamentals of Fixed Prosthodontics) and make some notes on it that you can easily discuss at your interview.

What Are the Interviewers Looking For?

Many applicants think that the most 'obvious' thing interviewers are looking for is excellent factual knowledge. This simply isn't true.

Interviewers are looking for an applicant that is **best** suited to study dentistry at **their** university. They are looking for your **motivation** to be a dentist. Standard dental school interviews generally don't test your innate recall knowledge of facts. Whilst having an excellent depth of knowledge may help you perform better during an interview- **you're very unlikely to be chosen based solely on your knowledge**.

Remember that the ultimate aim of this long selection process is to choose students who will go on to make good dentists in the years to come. When preparing for the interview and answering interview questions, it's very useful to think about how any of the qualities below could be shown in your answer.

Diligence
Dentistry is a very demanding profession and will require hard work throughout your whole career. Interviewers are looking not only for your ability to work hard (diligence) but also for an understanding that there will be times where you will have a responsibility to prioritise your dental work over other personal and social concerns (conscientiousness). This means for example that when you are a dentist with a free slot at the end of the day but your colleague is running behind, offer to see their last patient so you can both finish on time.

Professional Integrity
Dentistry is a profession where lives could be at risk if something goes wrong. Being honest and having strong moral principles is critical for dentists. The public trusts the dental profession and this can only be maintained if there is complete honesty between both parties. Interviewers need to see that you are an honest person, can accept your mistakes and are learn from them.

Empathy
Empathy is the ability to recognise and relate to other peoples' emotional needs. It is important that you understand and respond to how others may be feeling when you're a dentist. The easiest way to demonstrate this is by recalling situations from your work experience (one of the many reasons why it's so important). However, it's important to not exaggerate how much an incident has affected you - experienced interviewers will quickly pick up on anything that doesn't sound sincere.

Resilience to Stress
There is no denying the fact that dental school can be stressful due to the extreme pressure you'll be put under. Dental schools understand that you are likely to become stressed during your studies, however, it's important that you have a way of dealing with it in a healthy manner. Interviewers are looking to see if you are able to make logical decisions when put under pressure (something dentists encounter on a daily basis). Pursuing interests outside of dentistry, such as sport, music or drama is often a good way of de-stressing and gives you an opportunity to talk about your extra-curricular interests as well as team-working skills.

Scientific Aptitude
Dentistry is undoubtedly a science; whilst your grades will demonstrate your scientific aptitude, you'll need to offer something, in addition, to make yourself stand out. The easiest way to do this is by referencing examples of where you have gone beyond the confines of your school syllabus, e.g. Science Olympiads, Science Clubs, extra reading (hopefully you mentioned these in your personal statement).

Self-Awareness

You need to show that you have the maturity required to deal with complex issues that you will face as a dentist. The vast majority of students will undergo changes in personality throughout dental school and you need to have a good understanding of yourself to ensure these changes are positive ones.

Firstly, it's important to be able to recognise your own strengths and weaknesses. In addition, you need to be able to recognise and reflect on your mistakes so that you can learn from them for the future. A person with good self-awareness can work on their weakness to avoid mistakes from happening again. Questions like, "What are your strengths?" or "What is your biggest weakness?" are common and fantastic opportunities to let your maturity and personal insight shine through.

Manual dexterity

As a dentist you must be competent at using your hands with great precision and skill. The interview is a good opportunity to bring a sample of a piece of artwork or sewing, for example to demonstrate your capabilities. It is vital what ever you bring is your own work as you may be asked to do a live demonstration in the interview – such as a drawing. Additionally experienced interviewers will be able to quickly ascertain if you lying about work you claim to be your own.

Teamwork and Leadership

Working as a dentist requires working in a multidisciplinary team (MDT) – a group of individuals with a wide variety of skills that work to help patients. The MDT is a cornerstone of how modern healthcare functions. Thus, you need to be able to show that you're a team player. One of the best ways of doing this is by giving examples from your extracurricular activities, e.g. sports, music, duke of Edinburgh, other school projects.

> ***Top Tip!*** If you've got a bit of time before the interview then do something that shows your love and devotion to dentistry. For example, set up a mini dental journal club at your school where people can submit pieces on interesting aspects of science.

Realistic Expectations

Interviewers are looking for individuals that are committed to dentistry and have a realistic idea of what life as a dentist will entail. This is why it's extremely important to display your motivation to study dentistry clearly during the interview. A common way to do this is to reflect on your work experience and have examples ready of common challenges that you will likely face. Examples include:

➢ Long hours and stress
➢ Repetitiveness. A career in dentistry can sometimes demand great patience. Doing the same treatments day-in-day-out, seeing the same diseases, sitting through endless clinics
➢ The balance between being empathetic, yet remaining objective
➢ **Ethical dilemmas** – these will be discussed in detail in the 'Dental Ethics' section

Interviewer Styles

Although interviewers have wildly different styles, it is helpful to remember that none of them are trying to catch you out. They are there to help you. You may come across an interviewer that is very polite and 'noddy' while others may have a 'poker face'. Do not be put off by their expressions or reactions. Sometimes what you thought was a negative facial response to your reply may just be a twitch. Contrastingly, a very helpful appearing interviewer may lull you into a false sense of security. Rarely, you may get an interviewer who likes playing 'Devil's advocate' and will challenge your every statement. In these cases, it's important not to take things personally and avoid getting worked up.

You don't know what type of interviewer you will get so it's important to practice mock interviews with as many people as possible so that you're prepared for a wider variety of interviewer styles.

Communicating your Answers

Many applicants don't secure places as they don't spend enough time preparing for their interviews because they feel that they "already know what to say". Whilst this may be true, it is not always **what** you say but **how** you say it. Interviews are a great test of your communication skills and should be taken seriously. A good way to ensure you consistently deliver effective answers is to adhere to the principles below:

Keep it Short:

In general, most your responses should be approximately one to two minutes long. They should convey the important information but be focussed on the question and avoid rambling. Remember, you are providing a direct response to the question, not writing an English essay! With practice, you should be able to identify the main issues being asked, plan a structured response, and communicate them succinctly. It's important to practice your answers to common questions, e.g. *'Why Dentistry?'* so that you can start to get a feel for what is the correct response duration. Remember that MMIs are generally targeted so don't be shy to give slightly shorter (but still detailed) responses to them.

Give Examples:

Generic statements don't carry much weight without evidence to back them up. As a general rule, every statement that you make should be evidenced using examples. Consider the following statements:

Statement 1: "I am good at biology."
Statement 2: "I am good at biology as evidenced by my A* in biology AS and a gold medal in the UK biology Olympiad. I am an avid reader of 'Biology Review' as this allows me to keep up to date with new developments in the field."

Focus on Yourself:

It's not uncommon for candidates to go off on a tangent and start describing, for example, the many hurdles their young enterprise team faced. Remember, interviews are all about you - your skills, ability, and motivation to be a dentist. Thus, it's important that you spend as much time talking about yourself rather than others (unless absolutely necessary). Many students find it difficult to do this because they are afraid of being interpreted as 'arrogant' or 'a show-off'. Whilst this is definitely something to be aware of, it's important you do yourself justice and not undersell yourself.

Answer the Question:

This cannot be stressed enough – there are few things more frustrating than students that ignore the interviewer's questions. Remember, you need to **answer the question**; **don't answer the topic**. If a question consists of two parts – remember to answer both, e.g. *'Should we fluoridate water? Why?'*

Ending the Interview

At the end of the interview, the interviewers are likely to ask you if you have any questions for them. You should have gleaned enough information from the open days, school website, prospectus etc., that you do not have many questions. Unless the question is crucial, this may not be the right time to ask questions about the course, school or activities within the school as it will show your ignorance about the course/school. You can use this as an opportunity to show your interest in the course and dental school. This is also a chance to show that you have made the effort to research what makes their course 'special' to the interviewer. Perhaps show this by asking for more details e.g. *'I read that the course involves placements outside of the dental school, where are they based'?* If there is a current dental student you could involve them by asking, *'what do you think is the best thing about this dental school'* or *'Tell me about your elective?'* showing you know that that dental school has an elective period. However don't be perturbed that you do not have a question to ask.

MMIs may well end abruptly and you may be asked to stop when the end of time is signalled (bell rings or knock on the door). Avoid the temptation to linger on to answer the question that is half-answered unless you are asked to - extra time taken at one station means less time for the next one.

Dealing with Unexpected Questions

Although you can normally predict at least 90% of questions that you're likely to get asked, there will be some questions that you won't have prepared for. These assess your ability to handle difficult situations to see how you react to pressure and how you deal with the unknown.

Completely abstract questions are rare in traditional dental school interviews, it's unlucky to be asked a question that you don't have a clue about. If you do find yourself in this situation, do not panic. Pause and think. Have you come across something at school? Have you read something about it? Can you see/apply the basic principles? Can you these to start a discussion?

Good applicants will endeavour to engage with the topic and try to link in their knowledge to the question, e.g. *"I have not come across this before but it seems that…"*. Weaker applicants would be perturbed by the question or not consider it seriously, e.g. *"I don't know."*

Consider the example: *"Is there is life on Mars?"*

You may not have read about this specific topic. However, you may know about different space missions.

"As far as I know, there is not life on mars. However, given that there is some water on Mars, it would not be altogether impossible for there to be life."

Of course, this relies on you having a basal level of general knowledge. If you really have no clue at all and have not come across anything about this topic then it is better to acknowledge the gap in your knowledge.

"Unfortunately, I have limited information about this topic but it is an interesting question. I did read about NASA's mission to Mars but I'm unable to say if they have found life. I will endeavour to look this up."

Dealing with difficult questions like this gives you the opportunity to show that you are a motivated and enthusiastic student.

Interview Myths

1. Your interviewer is only interested in catching you out.
The interviewer is there to help you and to encourage you. It is best to think of the interview as a conversation between you and the interviewer. It is nothing more than an opportunity for you and the interviewer to get to know one another and see if you would be a good fit at that university.

2. The interview is the only important part of the admissions process.
Although the interview can be the final hurdle in some dental schools, many will also take your personal statement, academic grades and UKCAT/BMAT scores when making offers. Most dental schools will shortlist for interviews based on these so you cannot just focus purely on interview preparation from the outset.

3. You need to have immaculate knowledge of recent dental developments.
Many students claim that they are vicarious readers of Student BDJ and New Scientist on their personal statements, and then panic 'cram learn' the latest editions of these journals before their interview. Relax. You are not expected to be able to recite the articles from them. However, **you will be** expected to be familiar with current affairs and have a basic understanding of recent healthcare news.

4. All of the other interview candidates will know far more than you.
Whilst some students may give the impression that they have swallowed several encyclopaedias - don't let this put you off. Interviews are an opportunity to focus on yourself, expand your knowledge, and most importantly, show yourself as a potential dentist for tomorrow – not worry about whether you have read Shillingburg's Fundamentals of Fixed Prosthodontics back-to-back.

Structuring your Answers

You can approach questions in both MMIs and traditional interviews by using a simple framework:

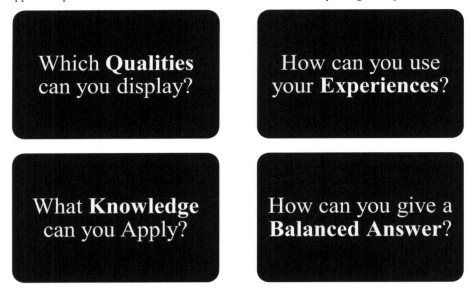

A good trick in MMIs can be to try and tick off each of these in an answer where applicable. This is a good way to ensure that you are not spending too long talking about one thing and so leaving yourself no time to talk about other things. There is little point in reeling off a list of unrelated facts when that would leave no time to show how your experiences apply.

It is better to demonstrate good knowledge and then move on to describe how insights from your work experience have also influenced you to give a more well-rounded answer. It's worth practising answering questions using this framework.

For example, consider: '*What do you think is the greatest challenge of being a Foundation dentist (FD)?*'

1. What knowledge can I apply?

I know Foundation dentists work in a similar way to an associate but they have a trainer in case they need additional support. I know they suddenly progress from a position of little responsibility as dental students to one of great responsibility and they have to make decisions on their own. I know the structure of dental training after dental school, which I might be able to work into the answer.

2. What experiences can I use?

I remember seeing a lot of dentists running late on my work experience. I remember one associate complaining that he was not getting his lab work back on time meaning he had to re book his patients. I remember the foundation dentist saying how scary it is to suddenly go from being a student to having a lot more responsibility and being expected to be able to practice relatively independently.

3. What positive qualities can I display?

I can show a realistic understanding of the challenges of dentistry. I can show a diligent and conscientious attitude towards the challenge of hard work and long hours. I can show an ability to handle stress.

4. How can I give a balanced answer?

I should recognise that there are multiple factors that make the foundation year a challenge in order to balance the answer.

You can go through a thought process like this in a few seconds and continue to think about it as you begin to answer the question. Thus, your answer may be along the lines of:

"There are clearly lots of different factors that are challenging for a foundation dentist. For a start, starting a job full time straight from university can be draining and scary but it is good to have the additional support of a trainer for that year. Though FDs will be used to hard work from dental school, this may still require a step up and even more sacrifices to be made in terms of a social life. However, whilst I was on my work experience I remember a foundation dentist telling me that the thing she found most challenging was the sudden jump in responsibility and expectations between being a dental student and being a Foundation dentist. It was this that she and her peers found most stressful."

You can see how you have to be very picky about what you include in your answer to make sure you give a well-rounded answer in the tight time limit of the MMI. For the same question, another response could focus on striking a work-life balance as being most challenging:

"Whilst doing my work experience, I saw foundation dentists working full time since graduating from university. In addition, they had to study for additional exams such as the MJDF to ensure their career progression. All this limited their social life. Thus, I think striking a good work-life balance could be challenging albeit possible with forward planning."

The STARR Framework

You may be asked questions where you need to give examples, e.g. *"Tell me about a time when you showed leadership?"*

It's very useful to **prepare examples in advance** for these types of questions as it's very difficult to generate them on the spot. Try to **prepare at least three examples** that you can use to answer a variety of questions. Generally, more complex examples can be used to demonstrate multiple skills, e.g. communication, leadership, team-working, etc.

In the initial stages, it's helpful to use a framework to structure your answers, e.g. the **STARR** Framework shown below:

Situation
- Explain the situation
- Explain answer to What, Where and When

Task
- Explain the Task
- Explain your Role and Responsibility

Action
- Explain what YOU did
- Briefly describe what sills you used

Result
- What was the result? Use figures when possible
- What difference did YOU make?

Reflection
- What went well? What did you learn?
- What would you do differently next time?

MULTIPLE MINI INTERVIEWS

A Multiple Mini Interview (MMI) is a series of short interviews. There are normally five to twelve stations and each station is five to ten minutes long. There is a new challenge and a new interviewer at each station. Most MMIs are one to two hours in total duration. All candidates must go through every station and will be asked the same questions. The major implications of this arrangement are:

1) You have a lot less time to make your points & build a rapport with the interviewer
2) There is a greater diversity of topics you can be tested on
3) You have a 'fresh start' at each station so can recover from poor previous stations
4) MMIs are fairer as everyone will get asked the same questions

The time pressure in MMIs makes communication challenging because there is less time to make a good impression on your interviewer and less time for them to judge you. This means that unlike traditional interviews, there is little time for ice-breakers; you'll need to give focussed yet comprehensive answers to ensure success.

What will I be asked?

Each MMI station will test different skills and topics like:

➢ One-To-One interviews on commonly asked topics, e.g. dental ethics, motivation to study dentistry, work experience, situational questions
➢ Writing Task where you're required to write answers to common questions. Normally, there are no interviewers
➢ Role Play where you need to interact with an actor, e.g. reassure a nervous patient
➢ Video Analysis where you need to comment on a video, e.g. of a dentist-patient consultation
➢ Rarely, there may be a rest station where there are sometimes drinks and snacks

A typical MMI circuit could involve:

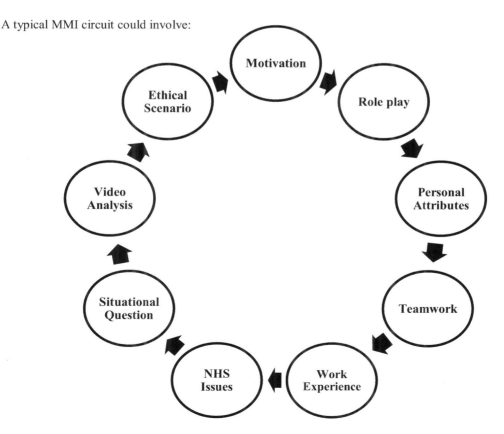

How to Tackle MMIs

Step One: Do your Research

Start off by finding out exactly how many stations there are and what duration. Then try to establish what topics are frequently asked and which topics have come up in the past. Are there any non-standard stations like maths or practical skills stations? Don't rely on your seniors – universities frequently change their MMI structures, so ensure you get the information from the university website or email the admissions office if you're unsure.

Step Two: Acquire the Knowledge

Once you have an idea about timings, you can start to tailor your preparation. For example, there is little point in practising extensive thirty-minute ethical arguments when the stations are only five minutes. Nevertheless, it's important to ensure you're familiar with the commonly asked questions. Again, don't spend extensive time trying to rote learn answers to common questions - this is not particularly helpful and is likely to be counter-productive.

Step Three: Practice with People

It's essential that you practice with a real person in advance. You need to practice giving succinct yet comprehensive answers as well as creating a strong first impression.

If there is more than one of you at your school applying for dentistry (or medicine), it would be worth considering that you could help each other and work together. Share resources, have practice sessions together and give each other constructive criticism. Even if you have only a couple of times a week when you're both free, it can really help to just sit down with someone and discuss an issue such as water fluoridation – each take a different stance, set a timer and see how you get on. Try not to be intimidated and keep calm. Five minutes might sound like a relatively short period of time, but it should be enough to get the most important points across.

Although many schools will arrange a practice panel or one-to-one interviews for you, very few will offer full MMI practice. If you feel you'd benefit from a full MMI circuit, consider attending the dental interview course which contains a mock MMI circuit (**www.uniadmissions.co.uk/dental -interview-course**).

> DON'T ASSUME THAT YOU WON'T GET ASKED TRADITIONAL STYLE INTERVIEW QUESTIONS IN MMIs. <u>YOU DEFINITELY WILL</u>!

Station 1: Role Plays

You may be asked to role play as a dental student or a foundation dentist. This is testing your ability to communicate information in a clear manner and your ability to handle stressful situations. You could be interacting with an actor or the examiner. You may need to talk to either a patient, family member or another dentist.

Examples Scenarios:

Your Role	Actor's Role	Location	Your Task
A Level Student	3rd Year Dental Student	University Open Day	Interact with the actor as you would at an open day
A Level Student	Your neighbour	The front drive	You accidentally run over your neighbour's dog with your car. Break the bad news to them
1st Year Dental Student	5th year dental student	On clinic	You notice that the 5th year dentist appears under the influence of ecstasy. What do you do?
1st Year Dental Student	Another 1st Year Student	Dental hospital	Your friend is thinking about leaving dental school as they don't enjoy it. Find out why and offer help
1st Year Dental Student	Patient	On clinic	Speak to the patient who has been waiting for over one hour and find out why they have come to the clinic
Foundation Dentist	Patient	In practice	Explain and apologise to the patient about a drug prescribing error that you contributed to
Foundation Dentist	Patient's Relative	In practice	The relative is very worried about their fathers oral health. Calm them down and answer their questions

Don't be scared by some of the horror stories you might hear where candidates were asked to break a diagnosis of cancer or tell a patient they were going to die. These are extremely unlikely scenarios that you won't be expected to handle (indeed many foundation dentists struggle with these). Even if you do get a particularly difficult scenario, try you remain calm- these stations are never a test of your knowledge. You are not expected to take a comprehensive history or diagnose a patient. **Role-plays are purely a test of communication.**

General Advice:

> Before starting the role play, ensure that you have read the scenario and understand exactly what is expected of you. Important things to double check are your role (dentist, student, etc.) and your location.

> Take a quick look around when you walk into the room. There are sometimes props positioned that you should utilise if appropriate. For example, you should offer a crying patient tissues or a cup of water.

> Always introduce yourself and your role unless the brief makes it abundantly clear that you don't need to (e.g. "Speak to your best friend Jane who is…"). A good opening line is: "Good morning, my name is Tom and I'm a first-year dental student. I'd like to speak to you about…"

> The actors are sometimes difficult and quiet on purpose in order to put you out of your comfort zone. Don't be thrown off if they appear to be reluctant to engage – that's their job!

> Don't be afraid of asking very vague questions to get the conversation started. Whilst questions like "How can I help?" or "How are you doing?" may appear redundant, they are important in opening a dialogue. Remember that going straight to the task can appear very forced, artificial and rude.

> Tailor your language depending on who you're speaking with. Be formal with patients and your senior colleagues; more informal with your friends.

> If you're uncertain about the direction the conversation is going, feel free to ask the actor: "Is there anything, in particular, you were expecting? Or anything I can do for you?"

> The principles of non-verbal communication still apply here. Maintain good eye contact, have an inviting posture and appear engaged in the conversation.

Station 2: Situational Judgment

The Situational Judgement station tests your ability to prioritise competing demands and resolve disputes in a harmonious manner.

If you have taken the UKCAT then you will have already come across SJT style questions. However, unlike the UKCAT, you may not get given multiple options to choose from. This means that you might have to generate the answer yourself rather than using the options as a guide.

In addition, this station may not have an interviewer. Instead, it may be a written station where you are expected to write your answer out in full rather than vocalise it. The scripts will then be marked collectively at the end.

When faced with a situational question, it is helpful to use a basic framework:
1. Identify the basic dilemma
2. Identify your potential courses of action
3. Consider the advantages & disadvantages of each option you're given
4. Consider any alternative solutions, i.e. options outside the ones you are given
5. Think about how the perspectives of colleagues and patients may differ from your own
6. Balance these to pick an action that you can justify with a logical argument

Treat Every Option Independently

The options may seem similar, but don't let the different options confuse you. Read each option as if it is a question of its own. It is important to know that responses should **NOT** be judged as though they are the **ONLY** thing you are to do. An answer should not be judged as inappropriate because it's incomplete, but only if there is some actual inappropriate action taking place.

For example, if a scenario says, "*A patient in the clinic complains she is in pain*", the response "ask the patient what is causing the problem" would be very appropriate. Any good response would also include documenting her complaint.

There might be multiple correct responses for each scenario, so don't feel you have to answer each of them differently. If you are unsure of the answer, mark the question and move on. **Avoid spending more than 90 seconds on any question**, otherwise, you will fall behind and risk not finishing.

Read "GDC Standards"

These are publications produced by the General Dental Council (GDC) which can be found on their website. The GDC regulate the dental profession to ensure that standards remain high. These publications can be found on their website and it outlines the standards that dentists are judged against and are expected to follow. Reading through this will get you into a professional way of thinking that will help you judge these questions accurately.

Step into Character

When doing this section, imagine you are in the scenarios being presented to you. Imagine yourself as a caring and conscientious dental student that you soon will be. What would you do? What do you think would be the right thing to do?

Hierarchy

The patient is of **primary importance**. All decisions that affect patient care should be made to benefit the patient. Of **secondary importance** are your work colleagues. So if there is no risk to patients, you should help out your colleagues and avoid doing anything that would undermine them or harm their reputation – but if doing so would bring detriment to any patient, then the patient's priorities come to the top. Finally, of **lowest importance** is yourself. You should avoid working outside hours and strive to further your education, but not at the expense of patients or your colleagues.

There are several core principles that you should attempt to apply to SJT questions that will help your decision making:

Adopt a Patient-Centred Approach to Care
This involves being able to treat patients as an individual and respecting their decisions. You should also respect a patient's right to confidentiality unless there is a significant risk to the general public. The most important principle is to **never compromise patient safety**.

Working Well in a Team
Teamwork is an essential part of any job. You must be a trustworthy and reliable team member and also communicate effectively within the team. You should support your senior and junior colleagues should they require it. It is important to avoid conflict and be able to de-escalate situations without jeopardising professional relationships where possible.

Commitment to Professionalism
You should always act with honesty and integrity as this is expected of anyone entering the profession. This includes apologising for your mistakes and trying to ensure other people apologise for theirs.

Taking Responsibility for your learning
Dentistry is a career where you are continuously learning. You are the sole person responsible for it and you will need to prioritise your jobs to ensure you attend scheduled teaching and courses. You should be able to critically reflect upon your experiences.

Example Questions

Example:
"The head nurse informs you that one of your colleagues is taking lots of supplies from the dental school home. What do you do?"

This question is not about your knowledge of dental materials but about what to do if someone you know is reported to be doing something that is not correct. The obvious trap here is to say, *"I would go speak to the dentist and inform the tutor."*

This would be a very serious accusation; remember that the nurse may be wrongly informed, biased or have a grudge against that dentist. Before you do anything, you must show that you will find out the facts and establish whether this is a recurring problem or not and if there are any obvious explainable reasons for it.

Thus, although the question appears very simple- it actually tests multiple skills. It's important to consider the implications of your actions rather than launch into an answer straight away.

"Taking dental supplies for personal use at home is something that worries me. Before I do anything, I would try to establish the facts by talking to others who may be in a position to observe such behaviour. This will ensure that it removes any reporter's personal bias or perceptions. If this were true, I would offer to speak to the colleague in private and ask their views. I would remind them of their professional responsibilities and I would encourage them to cease this behaviour and consider involving my seniors if I felt that the situation wasn't resolving."

Example B:
You are just finishing a busy emergency morning clinic. There is one new patient in pain left to see and your colleagues have already left to go to lunch. You need to leave on time as you have a social engagement to attend with your partner.

Rank the following actions in response to this situation in ascending order of appropriateness:

A. Finish off attending to the patients you have seen and take a brief history of the patient for the afternoon staff, then leave at the end of the clinic session on time.
B. Ask the patient to come back for the afternoon clinic and then leave at the end of your clinic time.
C. Make a list of the patient's problems to leave in an office for the afternoon staff to attend to.
D. Ask your tutor if you can leave the notes about the patients problems with them and then leave at the end of your clinic.
E. Leave a message for your partner explaining that you will be late and attend to the patient.

This question gives you the opportunity to demonstrate a conscientious attitude by prioritising patient care over personal concerns and team working. It would be unfair on your colleagues to leave the patient for them to see after lunch. The safest option is (E) as it ensures patient safety by helping a patient in pain and carrying out your duty of care. The other options all involve a delay of the patient being seen to and the history taking is done poorly and left for others to read which risks patient safety.

If you weren't given these options, you could instead talk the interviewer through your reasoning:
"So to summarise, I am faced with the situation where I am finishing my clinic and there is a patient in pain waiting to be seen. If I overstay, I will be late for my social commitment. If I leave, the patient is left waiting in pain and important information relevant to patient care may not get passed on.

In this situation, I have the option to leave a written history of the patients problem and either give them to the tutor on clinic or leave it in an office for the attention of the afternoon dentists. Whilst this may be reasonable, it's not the best option as there is a risk that the information may not get passed on at all or not in a timely manner which could compromise patient care. Or I could ask the patient to wait to be seen in the afternoon, this is unfair and potentially harmful to the patient who is in pain and shows a lack of professionalism.

I feel the best option would be for me to call my partner to let them know that I will be running late and provide the waiting patient with the necessary attention and treatment as required."

Many students find the Mnemonic ***INSIST*** helpful when structuring their answers:

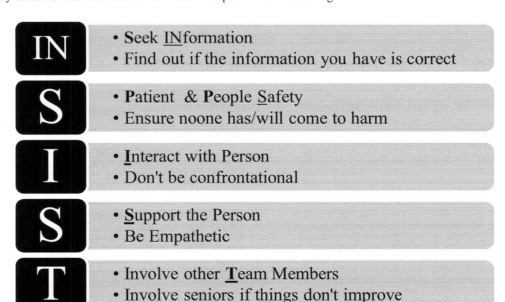

IN
• Seek <u>IN</u>formation
• Find out if the information you have is correct

S
• **P**atient & **P**eople **S**afety
• Ensure noone has/will come to harm

I
• <u>I</u>nteract with Person
• Don't be confrontational

S
• <u>S</u>upport the Person
• Be Empathetic

T
• Involve other <u>T</u>eam Members
• Involve seniors if things don't improve

Station 3: Problem-Solving Questions

You may be sent a problem-solving question a few days before your interview or get given it on the day before your interview is meant to start. These will usually not require detailed knowledge, but rather an ability to use your basic first principles. The important thing is that your argument is clear, logical, and you can give a logical step-by-step rationale for each step you made.

Consider: *A 70 kg man is trapped alive in an air-tight coffin of dimensions 2m x 0.5m x 0.3m. Estimate the amount of time that he has before he runs out of oxygen.*

➢ Clearly you do not have enough information to work this out very accurately, so you are going to have to make some assumptions. First of all, you need to work out the amount of air in the coffin, which requires you to estimate the amount of space being taken up by the man.

➢ If you use the facts that humans are made mainly out of water and that 1 gram of water = $1cm^3$ of water, you can estimate the man's volume as being 70,000 cm^3 =$0.07m^3$.

➢ This may not be very accurate, but it at least gives you a value to use for the rest of the calculation. If the volume of the coffin is $0.3m^3$, the volume of air in the coffin = 0.3 – 0.07 = $0.23m^3$ = 230 litres of air.

➢ As air is approximately 20% oxygen, you can estimate that there are about **46 litres of oxygen** in the coffin.

➢ If you assume that the man takes in about 0.5 litres of air with every breath, then he breathes in 0.1 litres of oxygen per breath. Now you have to assume the amount of that oxygen that he uses up before exhaling- say about 5%.

➢ So with every breath, he consumes 0.1 x 0.05 = 0.005 litres of oxygen. Therefore, if he can maintain a normal respiratory rate of about 12 breaths per minute, then he will consume 0.005 x 12 = **0.06 litres of oxygen per minute** and so the oxygen will last for$\frac{46}{0.06} = 766 \ minutes = 12 \ hours + 46 \ minutes$.

If you are given a question like this, you will be given a pen and paper to do your working. Try and lay out your working as clearly and methodically as possible to show that you have a clear line of thought. Additionally, write down your assumptions! This will help the interviewer follow your logic and will help you when you are justifying how you got to your answer.

If you are struggling to see how you would go about answering a problem, try looking at the numbers you have been given and think about what is there that is leading you, as you are unlikely to have been given irrelevant data.

If you can only see how to do part of the calculation then make sure you do that part anyway, even if it means estimating a number to use. You will at least get credit for being able to do some of it. However, do think about whether there is an alternative way to do it. If you are stuck, try to see which scientific principles you can apply.

Problem-solving questions can vary hugely. Sometimes, candidates can be handed dental instruments and be asked to use the properties of the instrument to deduce what it could be used for. The important thing to remember is that the interviewer is not expecting you to recognise the instrument right away, but rather wants to see how logical your thought processes are.

CURRENT ISSUES IN DENTISTRY

Fluoridation

Fluoride is a naturally occurring ion found in soils, water, plants and many different foods. It has been proven that Fluoride is beneficial to both developing and developed teeth in the reduction of tooth decay. Since the introduction of Fluoride toothpaste in 1914 there has been a significant reduction in dental caries. Dental decay is caused by bacteria found in plaque. Bacteria are able to produce acids from reacting with the sugars and carbohydrates that we consume. Fluoride acts in a number of ways. By incorporating into the hydroxyapatite crystals in enamel, it makes the teeth more resistant to attack from the acidic bacterial products. It will also promote remineralisation of the tooth structure from those weakened by the acids, repairing the teeth and can even arrest early enamel lesions. Finally studies have shown that fluoride can also interact with the metabolic pathway of bacteria resulting in a reduction in the amount of acids that are produced. While fluoride lowers the amount of dental decay, other factors are important such as good oral hygiene, salivary flow and frequency of sugar consumption.

Water fluoridation is the controlled addition of fluoride to a public water supply as a public health measure to reduce the incidence of dental decay. This is usually done at low levels of approximately 1 part per million (1ppm) sometimes written as one milligram per litre (1mg/l), as this is the optimum level required to protect against decay against harm caused. Currently around 10% of the UK population is fluoridated (around 6 million people)

The effect of fluoridation is drastic, especially amongst children. 5 year olds in areas with fluoridation schemes were less likely to experience tooth decay and were less likely to have severe decay than those not in fluoridated areas. This has a huge impact on oral health and important consequences as there is a significant difference in the numbers of children aged under 5 needing GA in fluoridated areas compared to areas with no water fluoridation. This is due to the fact the most common cause for GA in under 5s is dental extractions. While those in living in low social economic and deprived areas are 6 times more likely to have tooth decay, fluoridation of water benefitted both affluent and deprived areas but deprived areas had the most benefit.

The only proven side effect is dental fluorosis, which causes mottling of the teeth as there is a disruption of enamel formation while they are developing. This can range from mild white speckling to marked brown spots. In these mild cases it is only a minor aesthetic concern with no detriment to function and often can be easily treated with micro abrasion, resin infiltration or whitening.

Fluoridation has ethical issues, as it can be seen as mass medication and violating on a person's liberty and freedom of choice. In addition many consider fluorosis as causing harm to the population and the environment.

The Health and Social Care Act 2012 is the most up to date legislation regarding water fluoridation in England. It is down to the local authorities to decide the policies for water fluoridation and it requires public consultation. Usually the public need 3 months notice with proposals published in 2 consecutive weeks in the local newspaper. The local authorities must then decide in public and then it is the local water providers who are responsible for the addition of fluoride into the water supply. Trying to pass this motion is a great challenge as the water providers usually supply water to such a large area that it will range over more than one local authority. This means that every single local authority that is affected needs to approve to fluoridate the water otherwise it will not get passed.

The NHS Dental Contract

The current NHS dental contract has been in effect since 2006. It pays dentists based on the number of units of dental activity (UDA). Dentists earn UDA's by completing courses of treatments (COT) for patients. There are currently four bands, each one worth a different amount of UDA's. Remember to read '**What is included in each NHS dental band charge?'** on the NHS website for what treatments are covered in each.

Here is a simple breakdown of each band and how many UDA's each is worth:

➤ **Band 1 (1 UDA)**
   Diagnosis, treatment planning and maintenance. Examination, x-rays, scale and polish, preventative work, for example an assessment of a patient's oral health, minor changes to dentures.
➤ **Band 2 (3 UDA's)**
   Simple treatment, for example fillings (including root canal treatment), extractions and periodontal treatment.
➤ **Band 3 (12 UDA's)**
   Complex treatment that includes a lab element, for example bridges, crowns and dentures (excludes mouth guards).
➤ **Band 4 (1.2 UDA's)**:
   Urgent Treatment.

UDA's are awarded to the dentist for the completion of a COT. This means that once the band of treatment is established by the treatment required, it will be worth the same regardless of how many procedures are contained within it. For example a Band 2 COT with 3 restorations, an extraction and 1 root canal filling will be worth the same number of UDA's as a Band 2 COT with only 1 restoration.

It has been an area of controversy for many years now as it can be seen that the focus is put onto targets rather than patient care. Other problems with this current system are that dentists are target driven and are limited to how much time they can spend with a patient. Ideally the system should allow dentists to be more flexible and be able to give more time to those that need it. It also means that many NHS practices are reluctant to take on new patients if they have already reached their yearly targets.

The department of health is looking into a new reformed contract. Two different pilot schemes offering different ways to provide dental treatment have been trialled over the last few years with the focus being on prevention.

Use of Amalgam

Amalgam is a dental material that has been used since the 1800's and is silver in colour. It contains a combination of mercury, silver, copper, tin, zinc and other metals. It is known to be a hard wearing, durable and long lasting material that has been and still is (in many places) the material of choice for use in posterior teeth.

There has been a lot of controversy regarding this material especially over the last 30 years with a shift to phasing out this material in many countries. Countries in Europe such as Sweden, Norway and Denmark have now banned its use with many more likely to follow. The main controversy using this material has been its safety and the release of mercury, a toxic substance. However the World Health organisation (WHO) has recently published a statement to say that amalgam is safe to use. The main reason that we are now seeing this phase out is due the environmental disposal of this product.

Recently a new European Union Regulation 2017/852 on mercury has been published regarding the use and disposal of mercury. Article 10 of the regulation introduces provisions that need to be implemented by all dental professionals in the UK. It states that;

➢ By 01 January 2019 dental amalgam must only be used in pre dosed encapsulated form
➢ From 01 July 2018 dental amalgam must not be used for dental treatment of pregnant or breastfeeding women, children under 15 years, deciduous teeth (except when deemed strictly necessary by the dental practitioner based on the specific needs of the patient)
➢ From 01 January 2019 dental facilities must be equipped with an amalgam separator
➢ By 01 January 2021 all separators must retain at least 95% of the amalgam particles
➢ From 01 January 2018 dental practitioners must ensure amalgam waste is disposed of through an authorised waste management establishment

There are already alternatives to amalgam such as resin composites or glass ionomers. However while these alternatives are improving with time, there is still nothing that matches amalgam for wear and durability with a long-standing evidence base for its longevity. Amalgam also has the advantage of not being as moisture and technique sensitive unlike resin composites.

Sugar Tax

The Sugar Tax was first announced in 2016 and came into force on 6th April 2018. It is a tax on manufacturers, whose beverages contain sugars, meaning they will have to pay a levy on the high sugar drinks they sell calculated on the amount of sugar they contain. For example drinks with >8g per 100ml will face a tax rate equivalent to 24p per litre.

High sugar content beverages are a major risk factor for dental decay, obesity and diabetes. By introducing a tax on these beverages the aim is to deter people from buying them and raise money for the treatment of these ailments which costs the NHS millions every year.

Since the introduction opinion has been split. Many argue that it is a good thing as people buy and therefore consume less or companies will reduce the amount of sugar in their products. On the other hand some people are complaining that the government are interfering with their personal choice or that the tax does not cover high sugar foods such as cakes and sweets. Many companies have already reduced the amount of sugar with estimations of 50% of manufacturers have made changes since the tax introduction.

The UK is not the first country to introduce a sugar tax. Mexico introduced one in 2014 which has resulted in a decrease of 12% of high sugar beverage consumption. However the long-term effects are still up for debate.

Oral Cancer

Oral cancer or mouth cancer is rising significantly in both males and females, young and old. It is still more common in men but the numbers in females are rising more drastically. There are nearly 12,000 new cases in the UK every year with around 4000 deaths anually. Oral cancer can affect any part of the mouth and presents in a number of ways. The most common are persistent ulcers that do not resolve within several weeks, unknown causes for lumps that do not disappear or white/red patches that again do not resolve. Other signs are pain or difficulty swallowing, unexplained weight loss, bleeding or numbness in the mouth and unexplained loose teeth.

The prognosis for oral cancers remains poor. More than 6 in 10 head and neck cancers are diagnosed at late stages. One reason for this is that the public have a poor awareness of the signs and symptoms. In addition they are often quite aggressive and spread very quickly especially once they have reach the lymph nodes and glands. Incidence rates for oral cancer are projected to rise by around 33% between 2014 and 2035. Risk factors include lifestyle choices such as;

➢ Smoking and alcohol
➢ Diet - new evidence showing increased risk with consumption of red meat, poor nutrition
➢ Exposure to UV light
➢ Infections such as the Human Papilloma Virus (HPV)

Dentists are in a prime position for early detection of oral cancer as they can easily examine a patients intra-oral soft tissues at their regular dental check ups. Early diagnosis is key to getting swift treatment that will improve the prognosis of head and neck cancers. Dentists have a responsibility to be involved in all aspects of oral health and not just teeth and gums.

EXAMPLE INTERVIEW QUESTIONS

The list of questions below is not exhaustive but does cover a variety of question style that you are likely to be asked. However, there are no guarantees and you could potentially be asked things that you've never heard of anybody else being asked in interviews.

Nevertheless, don't be afraid of the unknown. Keep in mind the qualities that interviewers are looking for and go for it! Remember, what's more important is your ability to deliver genuine answers that are tailored to you – **don't try to rote learn and recite the example answers given in this book**.

1) Why Dentistry? Why do you want to be a dentist?

You are almost guaranteed to be asked this question in every dental school interview you attend. Hence it is essential that you have prepared a flawless answer for it. The best way to approach this question is by being very honest and very detailed in your answer. It can be difficult to know where to start when answering this question. You could look at your personal statement and bullet point the reasons you stated there of why you wanted to do dentistry or start a mind-map to help you reflect and drag out the deep-rooted motivation behind your decision to study dentistry. This could be the opening of your single long interview or a station in MMI.

Bad Response A:

"I have always been a caring person and interested in teeth. I cared greatly for my pets when I was younger and took a great interest whenever my friends or family became ill. Whenever I was taught a new scientific fact about the human body, I would always be the first in school to ask questions and do my own research to find out more. I have always admired dentists and looked up to them as role models in my life. There is no other profession that I respect more. In addition, my mum is a dentist and she'd like me to be a dentist too."

Bad Response B:

"I want to be a dentist because I would like to help people and work with my hands. I don't think I am made for an office job and dentistry offers lots of variety. Dentists play an important role in society which is something I would love to aspire to. The money's not bad either!"

Bad Response Analysis:

This answer shows very poor insight into the intricacies and the unglamorous reality of the dental profession. The candidate's example of their caring nature is distanced from the reality of using specialised dental skills and knowledge to treat patients. This answer is very focused on the past and does not mention how the candidate currently feels about dentistry. The candidate does not explain himself or herself in enough depth; they need to reflect more and explain why they respect the dental profession. The opportunity to show your qualities, i.e. hard work, keenness to learn, your empathic nature, is missed in the above response.

Each point has been poorly expanded on and leaves a lot to be desired. The answer is idealistic and does not demonstrate an understanding into what life is actually like as a dentist. Everyone says they want to help people and utilise their manual dexterity, so saying it in this way will not help you to stand out from the crowd. Although dentistry does offer a lot of variety, the way this answer is phrased makes dentistry seem like more of a last resort instead of something you have a burning enthusiasm for!

Dentists are indeed key members of society, but the answer fails to describe how and why this is, and instead, it seemingly focuses on the desire for prestige. It is perhaps best not to mention pay in your answer to this question, even if it is a factor for you!

Good Response A:

"I am very interested in people. I love working as part of a team and forming personal and professional relationships with others. I find it exceptionally rewarding to learn about how other people view the world and believe dentistry provides the opportunity to make uniquely trusting relationships between a patient and a dentist. I also have a great passion for science and scientific discovery. I am fascinated by scientific mechanisms and their complexities and I greatly enjoy the process of learning about them. My work experience has also consolidated my belief that a career in the dental profession is right for me. I was confronted with the unglamorous reality of dentistry for both the patient and the dentist, but I began to understand how rewarding the job of a dentist can be. I greatly admire the concept of making the care of the patient the dentist's first concern and using meticulous precision and skill to achieve the best results for my patients."

Good Response B:

"I found my volunteer work as a care assistant in X Hospice incredibly rewarding and this confirmed for me the decision to choose a career that would allow me to interact with a great variety of people and play a crucial role in their wellbeing."

Good Response C:

"There are many different reasons for why I think being a dentist is the right career path and lifestyle choice for me. First of all, I think dentistry as an academic discipline is absolutely fascinating and it is a genuine privilege to be able to even partly understand the way our oral health can affect our overall health. To be able to put this theory into practice in order to change lives would be the ultimate achievement. The skills obtained through studying and practicing dentistry are lifelong and transferrable throughout all walks of life, however, the ever-evolving nature of dentistry ensures that dentists are constantly learning new things and adapting their practices. These attributes of dentistry are incredibly important to me and I believe I would thrive under the pressure this would inevitably bring. I also believe I am a good communicator and would love a job that involves lots of interaction with other people: a vital part of being a dentist is interpersonal communication and teamwork, which I feel I excel at."

Good Response Analysis:

There is no one right answer; there should be lots of reasons for why you want to be a dentist! The above answer incorporates lots of different reasons to want to pursue dentistry – think about the qualities/attributes you think are important to being a dentist, and which ones reflect you personally.

Some of these mentioned above include: academia and hard work, practical skills, adaptability, communication and teamwork. It may sound cheesy/over-used to say you want to go into dentistry to help people, but it is actually true, so find a unique way of expressing this. Despite this answer being relatively short, all of these features are incorporated into it and demonstrate an insight into the life of a dentist. Moreover, I think it is also important to demonstrate an understanding that it is a lifestyle, more than just a job!

These answers are robust and complete. They show that the candidate understands that a wide range of skills are necessary to be a good dentist, that they have made a well-considered application to dentistry, and that they are passionate about a range of aspects of dentistry. References to a candidate's work experience is especially creditable as it shows an active interest in and a realistic perspective of dentistry.

Overall:

You must avoid a generic answer by making your response personal and including fine detail. Your answer to this question should be relatively long; make sure you highlight your passion for dentistry and use reflective language. You may find it useful to make some leading comments, which the interview can ask you follow up questions on; this can allow you to take control of the direction of the interview and show off your strengths. Be warned that your answer may be dissected by follow-up questions. For the above example, a potential follow-up question and a good potential answer are outlined below.

A good candidate can use the follow-up questions to give more details about their motivations and demonstrate even more of the qualities that the interviewers are looking for. Equally, if you give an interest in science and using science to help people as your primary motivation for studying dentistry, be prepared to explain why you would not rather go straight into a career in research. Similarly, think about what areas of human science have sparked your interest already. You can use what you have included in your personal statement to build your answer. For example, you could say "my project on the genetics of diabetes drew my attention to the relationship between diabetes and periodontal disease". Be prepared to have some discussion about your project, diabetes or periodontal disease or something completely different, e.g. public health issues and clinicians playing a role in disease prevention! It is important to be aware of what you have stated in your personal statement and read around relevant topics

2) What makes this university and its dental course right for you?

Do not confuse this question with 'Why do you want to go to dental school?'!

In your dentistry interview, you need to sell yourself to the interviewers and part of that is explaining why you'd suit the style of course that they offer. Try to research a little about the course before your interview and think about what attracted you to apply to that university and what it is about the course structure you particularly like. Also, try to highlight what you will bring to the course, particularly at universities offering a Problem-Based Learning style course where what the students themselves bring alters the teaching and dynamic considerably.

All universities are different, so it is important to have researched what the university has to offer that would benefit you and why/how you would fit in there. It is essential that the student knows what they are signing up for, and must show that they are the kind of person that would thrive in the sort of environment at that particular university. Thus, the answer to this question should make clear that they understand about the university and are the right sort of person for the challenge. Also, look into the extra-curricular activities and societies you might be interested in joining, e.g. your tennis or roving skills may be handy when applying to the university with an active tennis/roving team. Remember, the aim of the game is to portray how well you would fit into their university, and to demonstrate how you would contribute to university life. Sell yourself!

Having visited the university on an open day and chatted to some of the staff and students can be a good thing to talk about as it demonstrates your interest in the university enough to visit and your volition to seek out peoples' opinions. Obviously, the interviewers will assume you've applied to study dentistry at other institutions, but for this question a little ego massaging can go a long way.

Bad Response A:
"I've always wanted to study at Uni-X because it has such great facilities for both dentistry and for other aspects of student life. These facilities combined with the excellent teaching you have makes me think that Uni-X offers a course that would work well for me. I looked into the PBL-style course but I much prefer the idea of actually knowing something before starting to see patients, so that's why I've chosen a more traditional course like the one Uni-X offers."

Bad Response B:
"I want to go to this dental school because my interest in oral health and my desire to have a career where I can help people means I want to be a dentist. This dental school is in the top 5 in the league tables, so I think it would be good for my career to study here."

Bad Response Analysis:
This response sounds far too generic and does not give the interviewers the impression that you've visited the university or spoken to any current members. Try to explain why that specific university course is good for you and also why you would be good for that course. This response goes some way to doing that by saying that you would prefer a more traditional course. However, lots of places offer a more traditional course so make sure your answer is specific to that university. A poor candidate will often give an answer that does not properly answer the question, is not very specific to the dental school, and shows no/little consideration of the learning environment in the dental school, prioritising other factors disproportionately.

Good Response A:
"I understand that this university is extremely demanding and challenging, but this sort of pressurised environment is one in which I would thrive. I work well under pressure, and am also sure that the strong support networks that are established here will help me to complete my degree to the best of my ability. I am an active learner, and am always keen to be involved in teaching discussions. Therefore, I think that I would benefit greatly from the supervision system that is unique to this university. I am enthusiastic, sporty and outgoing, thus, would be keen to get involved with so much more that the university has to offer outside of my degree. I am very much an active person in all aspects of my life, which I think would make me fit in well amongst all the highly motivated students that come here."

Good Response B:

"The structure of the course at this dental school really appeals to me. I think I would get the most out of clinical experiences by already having a good understanding of the basic dental sciences, and so the split between pre-clinical and clinical studies here seems a great way to achieve that. Also, I am particularly interested in anatomy and know that this dental school offers some of the best opportunities for learning anatomy from cadaveric dissection, which seems the best way to learn it to me. Also, I like that there are opportunities for regular small group teaching at this dental school, so I am pushed to demonstrate my understanding of what I have been learning and get the opportunity to ask about anything I am unclear on. When I visited the school on the open day, I was particularly impressed by the choir."

Good Response Analysis:

This response demonstrates that the student is aware of the sort of learning environment that they would be in if they came here, and seems determined that they would do well under that sort of pressure. They also show knowledge of the supervision system that operates at this particular university, which is good because that is unique to this university. Additionally, the student considers life outside of work, which is good because the university is keen to have well- rounded individuals who will make the most of the opportunities there.

A good reply will be specific and will highlight the school's achievements, matching them to the applicant's own interest, linking what is said elsewhere to compliment this, e.g. personal statement, information in school's website or prospectus and own interest etc.

Overall:

Overall, to answer this question well it is essential to show knowledge of the university system and how your own attributes would make you suit that environment. It is important to show that you actually want to come to this university for reasons other than its prestige. The student must demonstrate that they have thoroughly researched the university and really do want to go there more than other universities, for well thought through reasons.

You need to give **specific** reasons for the dental school for which you are being interviewed. Answers like "this is the best school", or "my parents wanted me to attend the best school" may please the interviewer but you need to say something of substance. The dental school's website will usually give details of its course which you should familiarise yourself with and identify how it differs from courses at other dental schools. Show that the teaching style offered is right for your learning style and that the school offers excellent support to its students. A large difference among dental schools is between those that do the integrated course type (where clinical and theory-based teaching are given in parallel) and the split course type (those that give years of pre-clinical theory followed by years of clinical teaching). Candidates opting for the former often say it is because they cannot wait to get onto the wards and practice patient interaction throughout their time at dental school. Candidates opting for the latter often say that they want a solid grounding in basic dental science before entering the clinical sphere so that they can better understand their clinical experiences.

You may have visited the school on their open days so you could start your answer by adding something that you found interesting that matches with your style/personality, and if possible, link it with something in your personal statement. Following this, you can highlight the other equally important aspects of student life, e.g. social details of the dental school (societies, sports teams etc.), showing that you have put well-rounded consideration into your choice of dental school. So if the school has a good sports team or runs a drama club, you could link your sport/drama skills. A good candidate will demonstrate specific knowledge about the dental school, be positive about it, and have a matching personality/portfolio for choosing it.

3) What are the main principles of dental and medical ethics? Which one is most important?

There are a few principles which are generally accepted as the core of dental ethics. It can be useful to read a little about them to make sure you have a good grasp on what they mean. See the Dental Ethics section for more details.

In answering this question, good answers will not just state what the main principles of dental ethics are, but why they are so important. You don't need to go into too much detail, just show that you know what the principles mean. In choosing which principle is most important, it is good to show balanced reasoning in your answer, i.e. that you recognise that any single one of the principles could be argued to be the most important but for specific reasons you have picked this one.

A Bad Response:
"It is really important for dentists to empathise with their patients, so empathy is probably the main principle. If dentists empathise with their patients, this means they will do what is in the patients' best interests, so it is the most important principle in dental ethics."

Response Analysis:
Empathy is obviously an important component of the dentist-patient relationship and the ability to empathise is crucial for all dental professionals, but it is in itself not a principle of dental ethics. The candidate seems to be getting at the idea of beneficence but does not really explain what they mean or why it is so important. If you are asked to make a judgement on which dental principle is most important, it is a good idea to have mentioned other dental principles earlier in your answer to compare it to. Thus the main way in which this answer falls down is in having shown no real appreciation for so many of the dental principles mentioned above.

A Good Response:
"The main principles of dental ethics are usually said to be beneficence, non-maleficence, autonomy and justice. It is very difficult to say which is most important as, by definition, they are each crucial in dentistry and they are all very linked with each other. If I had to pick, I would say that beneficence is probably the most important principle as if it is applied to all your patients then it should imply justice and if a patient is properly respected, then it should also take into account the importance of autonomy."

Response Analysis:
This candidate showed knowledge in both the main principles and understanding of what they mean and how they could be applied practically. Notice how this candidate showed their understanding without actually defining each principle, though there wouldn't necessarily have been anything wrong with doing so. This candidate recognised the difficulty in picking a 'most important' principle, in so doing showing balance and humility. Their insight into the conflict between different principles further shows a good understanding and suggests they have given dental ethics a good degree of thought.

Overall:
This question can be a really easy one if you familiarise yourself with the main principles of dental ethics so that you know you can define them and recognise how they might apply in a clinical setting. The question of which is 'most important' has no single right answer – it can be good to say this and then make sure you have a reasonable justification for whichever principle you pick.

4) When can dentists break confidentiality?

As dentists, we have an ethical & legal obligation to protect patient information. This improves patient-dentist trust. However, there are circumstances under which dentists are allowed to break patient confidentiality like:

➢ The patient is very likely to cause harm to others, e.g. mental health disorder
➢ Patient doesn't have the capacity, e.g. infants
➢ Social Service input is required, e.g. child abuse
➢ At request of police, e.g. if a patient is suspected of terrorism
➢ Inability to safely operate a motor vehicle, e.g. had IV sedation

A Bad Response:
"Dentists can break confidentiality when they believe that it will benefit the patient. Confidentiality is an important aspect of dental care because it protects patients from having information that is private to them being disclosed to the public. Confidentiality can also immediately be broken if a patient who has been advised not to drive is found to be driving as it causes a risk to the public if the driver is involved in an accident."

Response Analysis:
The first sentence in this response is a little vague; ideally, the candidate would expand on how exactly breaking confidentiality would benefit the patient. For example, is it because the patient in question is in grave danger from abuse, threatened by a weapon or details about them need to be disclosed in order to catch the criminal? The question does, however, appreciate the importance of maintaining confidentiality. To state that confidentiality can be broken immediately without first consulting the patient (as in the driving scenario) and advising them of disclosing information themselves is technically inaccurate.

A Good Response:
"Patients have a right to expect their dentists to maintain confidentiality and doing so is very important to maintain a good patient-dentist relationship as well as maintain the public's trust in the profession. However, there are a number of circumstances under which dentists may break confidentiality. As stated in the GDC guidance, it may be broken if it is required by the law, if it is in the public's best interest due to a communicable disease such as HIV or a serious crime needing to be reported such as a gunshot wound. It is always important to ask patients first if their information can be disclosed and to encourage them to disclose things themselves. For example, if a person with uncontrolled epilepsy is driving, you have a duty to report it to the DVLA but you must give the patient every chance to tell the DVLA about their condition themselves. In this case, you can only break confidentiality if the patient refuses to inform the DVLA so that you can protect them and the general public."

Response Analysis:
The response identifies the importance of maintaining confidentiality. The candidate nicely references an appropriate source for dentists such as the GDC to state instances in which confidentiality is broken. This shows that the candidate has read guidance that dental students and dentists are expected to know and is able to apply her knowledge to answer this question. The examples are accurate.

Overall:
It is important to be familiar with key 'hot topics' in dentistry such as consent and confidentiality and to carry out a little background reading on these prior to attending a dental interview in order to provide them with accurate examples. Examiners will indeed be impressed if you understand these core concepts and their importance to dental practice. The GDC website is a very good site to visit in order to obtain further information on these topics and others that are likely to be assessed in your interviews.

5) How does NHS and private dental care differ?

Dentists in general practice may provide solely NHS or private dental care or a mixture of the two. The care provided in either domain must still follow the GDC guidelines and principles. However NHS care provided to patients must fulfil the criteria of clinically necessary. Therefore solely cosmetic treatments such as whitening are not available on the NHS. Additionally different labs and materials are used in private dentistry meaning the end cosmetic result may be superior as they are not restricted by an NHS budget when purchasing the latest materials and equipment. For example the lab cost of making a private denture is much higher compared to one on the NHS, as they will use much more natural looking teeth and acrylics.

The way patients pay for treatment also differs. Make sure you know the banding system for NHS dental treatments and how this differs to private FPI (fee per item) in which every treatment provided is charged individually. Look at the NHS choices website: **What is included in each NHS dental band charge?** For a detailed list of what things are included in each band of treatment.

NHS dentists are also limited on the time they can spend with patients due to the targets they are set to meet by the practice and the government. And unlike private dentists they are unlikely to work out of hours. However the higher cost of private treatment does not necessarily mean private dental care is better. There are no additional tests or qualifications required to become a private dentists.

When answering this question make sure you touch on each of the main differences between the two but try to avoid focusing on how dentists are paid. You should know about the UDA system (there is more about this in the 'current issues in dentistry' section at the end of this book) and how private associate are usually paid on an approximately 50-50 split of what the patient pays. However focusing too much on this tends to suggest you are more concerned with how much you are going to get paid rather than the care dentists provide.

A Bad Response:
I think the NHS dental care is not as good as the care that can be provided privately as the end results often don't look as natural. Also cosmetic treatments are not available on the NHS. However you have to pay more for private treatment, which is not an option for all patients.

Response Analysis:
This response shows little understanding of NHS or private dental care. Cosmetic treatments are available on the NHS, for example if a patient needed a crown on a front tooth to help restore function the NHS would cover a white crown to being made, it would not be made solely from metal – this is cosmetic. Only treatments, which are exclusively to improve aesthetics, are not available on the NHS. The response is also not very comprehensive as only two differences are mentioned.

A Good Response:
There are many ways in which private and NHS care differ. Private dentists may be able to employ best practice in regards to materials and the latest technology as they have little limitations on the purchasing of equipment. Private dentists are also able to provide treatment, which are not available on the NHS due to them not being clinically necessary such as whitening. I understand that NHS and private patients are not charged in the same way for their treatments as NHS patients are charged according to a banding system and Private patients are charged fee per item.

Although Private care may cost more and may look more natural this does not necessarily mean the care provided is superior to that of an NHS dentist. There are not any extra compulsory qualifications required to be a private dentist. Additionally due to target demands NHS dentists may be limited on the time they can spend with a patient, however this should not compromise their standard of care as both NHS and private dentists must adhere to the GDC guidelines.

Response Analysis:

A much more comprehensive response that analyses the different aspects of NHS and private dental care. Mentioning UDAs and GDC standards also opens the interviewer to asking more about these topics, which you should also be comfortable discussing.

Overall:

This question is not about just being able to list the main differences between NHS and private care. The interviewer expects you to understand the differences and then reflected on how patient care is affected. Be prepared to answer any questions regarding the benefits and limitations of both private and NHS care. Remember that this is an excellent opportunity to bring in your own experiences of working within these environments, which you will know from shadowing clinicians in your work experience.

6) What is a clinical trial?

Clinical trials are an important part of the licensing process of a new drug. There are several stages of clinical trials; all designed to assess safety and effectiveness of a drug. The clinical trial stage follows a series of other trials preceding it in order to provide an idea of safety for human testing. Before entering the clinical trial stage, a drug needs to be deemed safe for human use. To further ensure safety, reduced doses of the drug are usually used for early stage trials. Despite these steps, sometimes things go wrong in clinical trials and participants get hurt. Various institutions tightly regulate clinical trials.

A Bad Response:

"In a clinical trial, volunteers are given an experimental drug to determine its effectiveness in the treatment of a certain disease as well as its safety for use in humans. Human trials are superior to animal trials as they allow a judgement of the effectiveness of a drug in humans. After a drug has passed clinical trials, it is available to the general public in form of treatment. Through this process, we can progress quickly from drug formulation to use."

Response Analysis:

This answer is too short and shows little understanding of how dental licensing works. Whilst the general concept of clinical trials is correct, there are some errors regarding timescale as well as associated procedures and risks. Firstly, it takes about 10 years from the formulation of a drug to it entering the trial phase and even then, only a fraction of all drugs designed even enter the human trial phase. Secondly, there are several levels of clinical trials before a drug can be licensed for safe human use and even then, more time will go by on additional licensing procedures before it is available for use in treatments.

A Good Response:

"Clinical trials represent an important milestone in the licensing process of new medications. They usually are preceded by extensive testing on cells, computer models, and animals to provide an idea of effectiveness and safety in mammals. Human clinical trials themselves are organised in different stages, depending on how close a drug is to licensing. Phases 1 – 3 happen before a drug is licensed and involve increasing sample sizes as well as different parameters of research. Phase 1 trials, for example, involve the determination of safe doses, Phase 2 trials the determination of general effectiveness, and Phase 3 trials aim at comparing established treatments with new treatments to provide a judgement on superiority. It is important to note that not all drugs that enter Phase 1 trials will progress to Phase 3 and to licensing. Phase 4 happens after licensing and usually aims at determining long-term risks and benefits etc. Due to the nature of human testing, clinical trials of all phases are tightly regulated by governing bodies and the declaration of Helsinki as well as the Nuremberg Code."

Response Analysis:

This answer is a good one as it reflects a good knowledge of the process of clinical trials and also gives the examiners a quick summary of how precisely clinical trials work in the different phases. Showing insight into this is important as the term "clinical trial" is often misunderstood. It further demonstrates a good understanding of the challenges of human testing and the responsibilities arising from using human test subjects.

Overall:

Due to the use of human test subjects, clinical trials are an important ethical issue. As they play a central role in the development of drugs and the very real prospect of dentists treating patients that either are part of a trial or have an interest in participating in trials, it is essential for dentists to understand how clinical trials work and what differentiates the distinctive stages. Remember that all new drugs go through the same rigorous process before becoming available to the public.

7) Tell me about a time when you showed good communication skills

Communication is a very important part of a dentist's job. Most dentists work within teams and so will need to be able to effectively convey important information to many different members of staff who have different expertise. Communication is therefore essential for an effective integrated approach to patient care. Additionally, in order to ensure that the patient's autonomy is fully utilised, the patients themselves must be fully informed of all the details of treatment plans and procedures. This often requires excellent communication skills.

Bad Response A:
"I was the editor of my student newspaper and was responsible for ensuring the content of the paper was appropriate. I would often have to make guesses as to what the student body would like and then implement this into the paper. Often I felt that my writers didn't accurately convey what I wanted to be published in their articles, which was a real disappointment for me."

Bad Response B:
"Erm, once when I was working with a student I was tutoring, I had to explain how to calculate gradients since she wasn't understanding how the teacher was explaining it, so I sat her down and explained it to her. After going through it a couple of times, eventually she got it."

Response Analysis:
This candidate has identified a role they had where they potentially could have developed very sophisticated communication skills. However, the candidate suggests that they have not effectively communicated with their writers in order to get the most appropriate articles. Additionally, it would have been far more impressive had they said that, instead of guessing, they spoke with the student body to gain a consensus of opinions as to what the paper should be publishing. Instead, the candidate comes across as having approached the paper from an individual angle when good communication with different groups of students would have been more appropriate.

In the second response, the filler word "Erm" is poor form. It can be tough under pressure but avoid filler words. It's better to take your time and speak a bit slower if it means not using so much filler. Filling is not good communication! It gives off a tone of a lack of understanding. Moving on to the bulk of the response, the anecdote is a bit weak. The underlying narrative of helping a fellow student is a powerful one, but this response suffers because it lacks detail and it really sells the candidate short. Also, a note on the anecdote itself; saying things like "I sat her down" comes across as patronising. This is not the tone you want to project.

Good Response A:
"As part of my fundraising ventures, I organised a large ball to be hosted at the local town hall with a number of different aspects to the night including an auction, live musical performances and food. I needed to effectively coordinate my team so that everything ran smoothly. This entailed ensuring that there was sufficient press coverage of the event, arranging the logistics with the catering company and gathering goods to be auctioned. The night was a great success with plenty of money raised."

Good Response B:
"I currently work as a GCSE Maths tutor and often I have to explain difficult concepts to my students in a new way since their teachers' approaches often fall a bit short. An example where I feel I was particularly effective at communicating a concept was recently when I had to explain the idea behind gradients of straight lines. I tried to go at a pace appropriate for the student's ability and I kept an eye on the student's body language so I knew when I'd lost her and I could backtrack and isolate the bits she found most confusing."

Response Analysis:

This candidate has demonstrated an impressive ability to put on a major event, which has apparently entailed communicating with individuals from many different professions. The ability to maturely communicate effectively with so many different individuals to deliver a coherent and successful end product is very impressive. Despite this example being apparently far removed from dental practice, the candidate has effectively demonstrated that they hold fundamental communication skills that are widely applicable for the rest of their life, including their future dental career.

Response B is a considerably better response. Notice the lack of filler words. Here the key is also specificity. This is a good point to ensure that the anecdote you tell is actually true! Giving the specificity of the above response lends itself to belief. It's important to pull from real experience and the confidence of pulling from your actual experience will be evident in your own body language, and will yield a better outcome.

Beyond the details of the anecdote itself, this response is good because it actively highlights what the candidate actually did as far as good communication skills are concerned. The bad responses did not do this. The points about going at an appropriate pace and observing body language are absolutely essential to get across in your response because it shows that you actually know what the interviewer means by good communication skills.

Overall:

Whatever the example you use, make sure that it shows you have developed or are developing good communication skills. At the end of your example, it would then be useful if you could demonstrate an awareness of why communication in dentistry is so important, e.g. communicating tests results to the patient or breaking bad news (diagnosis of cancer).

Ultimately, this question is all about the details. Both the good and bad responses used similar anecdotes, but the good response has the level of detail necessary to succeed. There is, of course, a balancing act between giving enough detail to sound reliable and going so far as to become esoteric and alienate the interviewer. Crucially, it's important to highlight what actually counts as good communication skills. Don't fall into the trap of just assuming that both you and the interviewer know what good communication skills are. It's a nebulous term, so going out of your way to specify what qualities it includes is a valuable thing to do.

8)   How would you tell a patient they've got oral cancer?

This is a very testing question; it tests your ability to cope well under stress. It is important that you do not panic. When you enter an interview, you should be expecting the unexpected and be broadly prepared for everything and anything! To answer this question well you will need to think logically and work through your answer step-by-step. You are not expected to know any formal protocol on how to deliver bad news to a patient. You should be aware that the interviewer will be looking to see evidence of your empathy skills, your problem-solving skills and your ability to think on the spot. A good answer will be based around the keyword in the question – "tell"; i.e. how will you articulate, communicate or disclose this information?

A Bad Response:

"I would ask the patient to sit down. I would then use a very serious tone of voice. I would try not to show any emotion in my facial expression and be very factual with the information I was passing onto the patient. I would not keep the patient in suspense; I would tell them as soon and as quickly as possible."

Response Analysis:

This answer is too short and not enough thought has been put into it. The candidate has not explained the intended impact of his actions of the patient. If he had, he may have realised that he would have come across as rather cold. The answer also suggests that the candidate would rush this interaction with the patient rather than taking his/her time to deliver the news. The candidate has not identified why the information he is disclosing to the patient is important and seemingly skirts around the fact that he is revealing to a patient that he/she has a terminal illness. The interviewer needs to know the candidate is able to talk about difficult issues in a mature and sensitive manner. The answer above does not demonstrate this.

A Good Response:

"I would be very aware of the patient's current position; the level of awareness and understanding the patient currently has about his condition, available sources of support for this particular patient – for instance, family and friends, religious circles, specific support groups – and the current anxiety and distress the patient is feeling as a result of their poor health. I would adapt the specifics of what I would say as appropriate for an individual patient. But most fundamentally, I would use clear and simple communication and a very professional manner. I would give a thorough background explanation of what has led the dentists to reach this conclusion. I would check throughout the conversation that the patient has understood what I have said before I move onto the next point. I would openly show my empathy and sympathy to the patient by letting the patient know that I am terribly sorry that they are in this situation and by offering my support and patience, and inviting them to ask any questions they have."

Response Analysis:

The candidate shows exceptional empathy skills; they seek to understand more about the patient to understand more about how they will feel and how they can potentially cope with the information they are about to receive. The candidate understands that this consultation is a very personal and sensitive interaction between patient and dentist. This response shows awareness of the importance of good quality communication and understanding of context to help the patient make sense of what is helping them and how they can cope with this life-changing event.

Overall:

The interviewer will want to see that you realise you are doing much more than simply communicating a fact to a patient. They will want to see that you understand the significance of this information to the patient and hence the significance of the style of (well-informed) communication you use.

9) *You are a dental student on clinic. Your tutor in charge turns up on Monday morning smelling strongly of alcohol. What do you do?*

Situational judgement questions aim to assess your approach to complex scenarios which you may encounter in your workplace. They are designed to test your potential across a number of competencies. In this case, this question tests your ability to deal with a senior colleague whom you suspect is drinking alcohol and has turned up to work smelling of alcohol. There is usually a pattern to follow when answering these questions: try to approach the person in question to gather a bit of information- are they, in fact, drinking alcohol? You may have been mistaken and it would, therefore, be wrong to take any further action. Next, you should try and explore the reason behind their behaviour- is it a transient and short-lasting event that has caused the consultant to drink? If so, hopefully there shouldn't be a long-term issue here. Thirdly, the interviewer would like to hear that you are taking steps to ensure that patients are safe. This may involve asking the tutor politely to get some rest and go home- clinical errors or prescribing errors due to alcohol consumption is dangerous. Lastly, you may want to suggest the consultant seek some help.

A Bad Response:
"Smelling strongly of alcohol at your workplace is, in my opinion, unacceptable. The tutor, although a senior figure, should know better and I think his behaviour should be reported promptly. The consequences of having a drunken tutor in the clinical area are unsafe and it also tarnishes the dentists' reputation as a whole. I would therefore ask another tutor to have a word with the tutor and hopefully, the matter will be escalated to the Dean of the dental school who can then decide the best course of action."

Response Analysis:
This candidate is rather rash in his/her approach to the situation. Firstly, there is only a suspicion that the tutor is drinking alcohol- it is thus better to sensitively explore this first before discussing the situation with anybody else. Reporting somebody without first getting the facts straight is inappropriate. Lastly, whilst you can seek help from another tutor who will be senior to you, the answer here sounds more like you are passing the buck to the tutor and asking him to sort the situation out rather than seeking advice and acting on the advice yourself; interviewers will appreciate you being proactive and sorting matters out yourself.

A Good Response:
"This is a complex scenario. As there is only a presumption here that the tutor has been drinking (he smells of alcohol only), I would tentatively approach him and politely ask him if he has been drinking any alcohol. I would next offer to explore his behaviour by asking him what has led him to drink alcohol and be so out of control that he still smells of it when he comes into work. I would then suggest he takes the rest of the day off after ensuring his shift is covered by explaining that patient safety may be compromised if he practices dentistry under the influence. Lastly, if I believe that this may be a long-term problem, I would suggest to him that he seeks further help, either by going to his GP or going to occupational health."

Response Analysis:
This answer takes a calm and measured approach to the situation by following the 'usual' steps for this type of scenario. The candidate is information gathering rather than reporting the consultant straight away. There is also an awareness that patient safety may be at risk, and the candidate provides a solution to tackle this and understands the need to be sensitive here.

Overall:
Situational judgment questions will be difficult and the key is to take a measured and calm approach to the situation. Reporting individuals straight away before attempting to resolve the situation between teams is often not the right approach, but interviewers would rather you to talk to the person in question yourself and take it from there. But remember that patient safety is the most important aspect here and if the tutor were to refuse to go home and continue seeing patients under the influence, you may then need to escalate the situation to someone more senior to you to ensure that patients are not in danger.

10) Can you describe a time you showed leadership?

Leadership is a key quality in a dentist. The important thing is to show that with this experience you had to demonstrate qualities such as initiative, decisiveness, organisational abilities and the ability to manage, guide and motivate others. These are skills that are often used in a dental setting.

It may be that you have not had any leadership roles, however, these experiences can be small. If you have ever taken the lead in organising a social event or group activity – a party, group trip, book club – then this can be used as an example. Describe the situation, how you came to be in a leading role, the steps you took to keep things running smoothly, and the result – for example, a successful event or crisis averted.

When answering this question, it is helpful to think of good qualities in a leader and weave them into your answer. Remember that managing a team doesn't necessarily make you a great leader. It's important to choose a story that demonstrates true leadership — stepping up to guide or motivate or take initiative, ideally in challenging circumstances. Do not be afraid to sell yourself!

Bad Response A:
➤ *"I do not often take on the leadership role. I prefer to work as a team member and work together to achieve a collective goal."*
➤ *"I'm never in charge so I never really get an opportunity to show leadership."*

Response Analysis:
Although it is essential that dentists act as team members, they also pay a dual role as team leaders due to the multidisciplinary nature of healthcare. Both answers don't give any example of leadership and are thus poor. It is clear that the candidates don't realise the importance of leadership in the role of a dentist.

Good Response A:
"I am a Charity Officer currently and had to organise the Charity Fair. I scheduled a meeting for the full team to discuss ideas but discovered that four people had dropped out. We had to divide the responsibilities between the remaining team members as a result, people were overworked and morale suffered. I baked for each weekly meeting to demonstrate my appreciation for all of their hard work during a challenging time. I also ensured that I asked them for ideas on how to be more efficient. I made it clear that no idea was stupid and that it was a safe environment for making suggestions. Performance improved and we all worked together to make the Charity Fair a huge success."

Good Response B:
"After having a string of lost games in my school rugby team, I noticed that morale in the team was getting low. To try and counteract this, I tried to lead by example and be as positive as possible and encourage my team-mates to come to training and make sure they knew exactly when and where it would be. I think this was important in stopping some of my team-mates from choosing to drop out of the team and was one of the reasons why we got a win soon after. I know it is only rugby, but I think being able to help organise and motivate a team could be relevant to dentistry."

Response Analysis:
The answer shows many key qualities needed in a leader. By baking, the candidate shows that they understand the importance of motivating team members and showing appreciation for their work. The candidate also shows great listening and communication skills.

Overall:
A good answer will demonstrate knowledge of the qualities seen in a good leader such as communication skills, motivating others and inspiring trust. A better answer may relate the skills shown to what they have seen in clinical practice.

11) What is your biggest weakness?

This question often catches candidates out. A careful balance needs to be struck between not coming across as arrogant and demonstrating a dangerous lack of insight (nobody is perfect!). It is also important not to identify a weakness that may make the examiners worry about who they are admitting to read dentistry. The answer needs to identify a weakness and then demonstrate how you are attempting to rectify this.

Avoid the urge to give an answer that is not really a weakness, e.g. "I'm a perfectionist", as a more profound degree of self-criticism will be appreciated more. Instead, by simply changing the phrasing and customising it to yourself, you can say something very similar, e.g. *"my time management can be a weakness, often because I spend too much time concentrating on minor details in my work"*. You would need to develop this by showing what you've learnt from reflecting on your weakness, e.g. *"...so when I am working on a project, I set myself strict targets to meet and try to get the bulk of the task done before worrying about the finer details in the early stages."*.

Showing humility when answering a question such as this is a good way to make you appeal to the interviewer as a person. Some candidates may be tempted to answer this with a joke, however, this is not advisable as it is a very important question for medics.

A Bad Response:
"My biggest weakness is that I'm sometimes quite arrogant. I've always been very successful at most of the things I try and do and so have an awful lot of self-belief. I rarely feel the need to ask for help as I know that I will likely be OK without it. When I fail at something, I get very angry with myself and will often feel down for several days afterwards."

Response Analysis:
This is a dangerous response. Many dental applicants will indeed be very good at lots of things; however, saying that this has made them arrogant is not good. The candidate states that they rarely seek help for things and have an innate belief that they will be good at anything they turn their hand to. It is not possible to be good at everything from the offset, and knowing when to ask for more senior help is an absolutely critical part of being a dentist. In dentistry, it's not just that your ego will be bruised if you fail, but a patient's health may be severely affected.

A Good Response:
"I consider myself a natural leader and so my instinct in many situations is quite often to lead the team and exert my authority in coordinating others. However, I appreciate that quite often during my dental career I will not be the most appropriate individual to lead a team. Therefore, I'm working really hard on my skills in working effectively within a team and not necessarily as the leader. In order to improve on this, I've recently started working at my local carpentry shop with little experience, and am therefore very reliant on senior advice."

Response Analysis:
This candidate identifies a valid weakness that, if too extreme, would be a problem in dental practice. However, crucially the candidate has demonstrated insight into this problem and how this could be an issue as a future dentist. Impressively, they've even shown that they are being pro-active in trying to address this weakness (working as part of a carpentry team).

Overall:
Dental practice is quite rightly very keen on quality control and ensuring best practice. Therefore, it is essential that those that practise dentistry have sufficient insight to identify where they are struggling and when to ask for help. Be this a surgeon whose struggling with a new procedure or a dentist who fails to take into account the opinions of his team sufficiently. A good dentist will identify their flaws and adjust these so as to prevent the quality of patient care being affected.

12) How did your work experience change your views of dentistry?

This question is essentially giving you an excellent opportunity to present concisely what you learnt from your work experience. There is a need for you to compare your views on dentistry before and after the work experience, which is potentially tricky. However, if answered correctly, this question is inherently structured in a way that allows the candidate to easily make the critical observations interviewers look for when talking about work experience. Namely, that the candidate has <u>learnt</u> something from the W/E rather than just observed passively. Therefore, the candidate should aim to identify a particular situation they observed during W/E and how this challenged their preconceptions about dentistry, and subsequently usefully informed them about the career. This is how W/E questions should generally be approached, as opposed to merely reeling off a list of things that you saw.

A Bad Response:
"Whilst observing a surgical removal of a lower wisdom tooth, I was shocked by the amazing skill of the surgeons in creating a flap to uncover the bone to aid in extracting the tooth. I was impressed how he managed to suture the gums back in place in such a restricted environment.

Response Analysis:
Identifying the "amazing skill" as something that you've learnt is probably inadvisable. It's not very specific, and it shouldn't really be that "shocking" if you're seriously considering a career in dentistry. Additionally, it wastes a good opportunity to talk about something far more insightful and useful to a career in dentistry. The second issue with this answer is the very simplistic explanation of what the surgeons were doing during the procedure. It suggests that perhaps you didn't really know what they were doing and didn't <u>actively</u> participate in your learning, and rather were merely passively by-standing.

A Good Response:
"I observed a surgical removal of a lower wisdom tooth and asked the surgeon beforehand about his approach. I was fascinated to learn about how the surgeon assesses the tooth first in order to assess the difficulty and risks involved in removing it. We also spoke about and his flap design which he wanted to keep minimal to aid in recovery. I was really able to appreciate the skill required to work in such a tight space right at the back of the mouth. This also allowed me to observe not just the oral surgery side of dentistry but also how the surgeon managed this patient who was understandably a little anxious about the procedure. Communication was key in making the patient feel more comfortable and reassured. Previously I'd seen dentistry as restoring and removing teeth but this experience has made me realise the patient's perspective to dentistry and that being capable of managing them effectively is just as important as clinical skill.

Response Analysis:
This response identifies two key learning points for the candidate: the more general observation of the importance of having manual dexterity to deliver complex dental treatment and also the patient management required. This is ideal as it shows both an interest in the dynamics of dental practice as well as the underlying science. Importantly, the candidate sticks to the question asked and emphasised that prior to his work experience, he didn't appreciate the importance of how important patient management is. He also leaves plenty of "carrots" for future questions from the interviewers e.g. what are the risks of removing lower wisdom teeth? What features makes a tooth more difficult to extract?

Overall:
When answering questions about "what you learnt from your work experience", you should always aim to show how your views/understanding of dentistry has been changed by the experience. This is instead of simply reeling off lists of what you've seen- it's all about quality and not quantity. Organise your thoughts before the interview and categorise different experiences into different learning categories. This will help you to present your thoughts as efficiently as possible.

13) Which of the two molecules below is more acidic? What factors make this the case?

(A) $—OH$ vs.   (B)

This question is introducing the candidate to the idea that the **concept of acidity** can be applied to more molecules than just the classic "acids" you learn at school.

A good candidate would first define acidity:

$$HA \rightleftharpoons H^+ + A^-$$

Then need to **highlight the key reactive areas** on each of the molecules and assign how each of the molecules would behave when behaving as an acid. In this case, both molecules form $RO^- + H^+$ as the products. The crux of this problem is that the stability of MeO^- is greater than $(Me)_3CO^-$ which is because the O^- is more stable in A.

Methyl groups are electron donating groups and in molecule (B) there are three Me groups pushing onto the carbon bonded to the oxygen, therefore, this carbon is more electron rich than molecule (A) so destabilises the O^-. Therefore, the equilibrium for molecule (B) in water is more shifted towards ROH rather than RO^- so molecule A is more acidic than molecule B. A good candidate will also then link this to equilibrium constants.

$$K_a = \frac{[H_3O^+]_{eq}[A^-]_{eq}}{[HA]_{eq}}$$

This question should not be too difficult - good students would be expected to give a comprehensive answer that synthesises multiple chemistry principles from the A-Level syllabus. This question tests how comfortable people are with these principles and if they can use them in different scenarios.

Final Interview Advice

Some DOs:

- ✓ **DO** speak freely about what you are thinking and ask for clarifications
- ✓ **DO** take suggestions and listen for pointers from your interviewer
- ✓ **DO** try your best to get to the answer
- ✓ **DO** have confidence in yourself and the abilities that got you this far
- ✓ **DO** a dress rehearsal beforehand so that you can identify any clothing issues before the big day
- ✓ **DO** make many suggestions and have many ideas
- ✓ **DO** take your time in answering to make sure your words come out right
- ✓ **DO** be polite and honest with your interviewer
- ✓ **DO** prepare your answers by thinking about the questions above
- ✓ **DO** answer the question the interviewer asked
- ✓ **DO** think about strengths/experiences you may wish to highlight
- ✓ **DO** remember to bring examples of manual dexterity

Some DON'Ts:

- ✗ **DON'T** be quiet – even if you can't answer a question, how you approach the question could show the interviewer what they want to see
- ✗ **DON'T** be afraid to pause for a moment to gather your thoughts before answering a question. It shows confidence and will lead to a clearer answer
- ✗ **DON'T** give them attitude or the feeling you don't want to be there
- ✗ **DON'T** rehearse scripted answers to be regurgitated
- ✗ **DON'T** answer the question you wanted them to ask – answer the one that they did!
- ✗ **DON'T** lie about things you have read/done (and if you already lied in your personal statement, then read/do them before the interview!)

Interview Day

- ➢ Get a good night's sleep
- ➢ Take a shower in the morning and dress at least smart-casual. It is probably safest to turn up in a suit
- ➢ Get there early so you aren't late or stressed out before the interview even starts
- ➢ Don't worry about other candidates; be nice of course, but you are there for you. Their impressions of how their interviews went have nothing to do with what the interviewers thought or how yours will go
- ➢ It's okay to be nervous – they know you're nervous and understand, but try to move past it and be in the moment to get the most out of the experience
- ➢ Talk slowly and purposefully; avoid slang and not use expletives.
- ➢ Try to convey to the interviewer that you are enjoying the interview
- ➢ It is very difficult to predict how an interview has gone so don't be discouraged if it feels like one interview didn't go well – you may have shown the interviewers exactly what they wanted to see even if it wasn't what you wanted to see. Indeed, many people who are given an offer after their interview had felt that it had not gone well at all
- ➢ Once the interview is over, take a well-deserved rest and enjoy the fact that there's nothing left to do
- ➢ Above all, smile and enjoy the experience!

Afterword

Remember that the route to success is your approach and practice. With targeted preparation and focused reading, you can dramatically boost your chances of getting that dream offer.

Work hard, never give up, and do yourself justice.

Good luck!

Acknowledgements

We wish to thank the many tutors and colleagues for their help with compiling this mammoth book – it wouldn't have been possible without you all. I'm hopeful that students will continue to benefit from your wisdom for many years to come.

About *UniAdmissions*

UniAdmissions is an educational consultancy that specialises in supporting **applications to Medical School, Dental School and to Oxbridge**.

Every year, we work with hundreds of applicants and schools across the UK. From free resources to our *Ultimate Guide Books* and from intensive courses to bespoke individual tuition – with a team of **300 Expert Tutors** and a proven track record, it's easy to see why *UniAdmissions* is the **UK's number one admissions company**.

To find out more about our support like intensive **courses** and **tuition**, check out **www.uniadmissions.co.uk**

YOUR FREE BOOK

Thanks for purchasing this Ultimate Guide Book. Readers like you have the power to make or break a book – hopefully you found this one useful and informative. If you have time, *UniAdmissions* would love to hear about your experiences with this book.

As thanks for your time we'll send you another ebook from our Ultimate Guide series absolutely <u>FREE</u>!

How to Redeem Your Free Ebook in 3 Easy Steps

1) Either scan the QR code or find the book you have on your Amazon purchase history or email your receipt to help find the book on Amazon.

2) On the product page at the Customer Reviews area, click on 'Write a customer review.' Write your review and post it! Copy the review page or take a screen shot of the review you have left.

1) Head over to www.uniadmissions.co.uk/free-book and select your chosen free ebook! You can choose from:

- ✓ BMAT Mock Papers
- ✓ BMAT Past Paper Solutions
- ✓ The Ultimate Oxbridge Interview Guide
- ✓ The Ultimate UCAS Personal Statement Guide
- ✓ The Ultimate BMAT Guide – 800 Practice Questions

Your ebook will then be emailed to you – it's as simple as that!

Alternatively,   you   can   buy   all   the   above   titles   at

www.uniadmissions.co.uk/our-books

BMAT ONLINE COURSE

If you're looking to improve your BMAT score in a short space of time, our **BMAT Online Course** is perfect for you. The BMAT Online Course offers all the content of a traditional course in a single easy-to-use online package- available instantly after checkout. The online videos are just like the classroom course, ready to watch and re-watch at home or on the go and all with our expert Oxbridge tuition and advice.

You'll get full access to all of our BMAT resources including:

✓ Copy of our acclaimed book "The Ultimate BMAT Guide"
✓ Full access to extensive BMAT online resources including:
✓ 10 hours of BMAT on-demand lectures
✓ 8 complete mock papers
✓ 800 practice questions
✓ Fully worked solutions for all BMAT past papers since 2003
✓ Ongoing Tutor Support until Test date – never be alone again.

The course is normally £99 but you can get **£ 20 off** by using the code *"UAONLINE20"* at checkout.

https://www.uniadmissions.co.uk/product/bmat-online-course/

£20 VOUCHER: UAONLINE20

MEDICAL INTERVIEW ONLINE COURSE

If you've got an upcoming interview for medical school but unable to attend our intensive interview course– this is the perfect **Medical Interview Online Course** for you. The Online Course has:

✓ 40 medical interview on-demand videos covering Oxbridge and MMI-style questions.
✓ Copy of the book "The Ultimate Medical Interview Guide."
✓ Over 150 past interview questions and answers.
✓ Ongoing Tutor Support until your interview – never be alone again

The online course is normally £99 but you can get £20 off by using the code "*UAONLINE20*" at checkout.

https://www.uniadmissions.co.uk/product/online-medical-interview-course/

£20 VOUCHER:
UAONLINE20

UKCAT ONLINE COURSE

If you're looking to improve your UKCAT score in a short space of time, our **UKCAT Online Course** is perfect for you. The UKCAT Online Course offers all the content of a traditional course in a single easy-to-use online package- available instantly after checkout. The online videos are just like the classroom course, ready to watch and re-watch at home or on the go and all with our expert Oxbridge tuition and advice.

You'll get full access to all of our UKCAT resources including:

✓ Copy of our acclaimed book "The Ultimate UKCAT Guide"
✓ Full access to extensive UKCAT online resources including:
✓ 10 hours of UKCAT on-demand lectures
✓ 6 complete mock papers
✓ 1250 practice questions
✓ Fully worked solutions for all UKCAT past papers since 2003
✓ Ongoing Tutor Support until Test date – never be alone again.

The course is normally £99 but you can get **£ 20 off** by using the code *"UAONLINE20"* at checkout.

https://www.uniadmissions.co.uk/product/ukcat-online-course/

**£20 VOUCHER:
UAONLINE20**

Printed in Germany
by Amazon Distribution
GmbH, Leipzig